Qiang Lu, Ying Chen, Xuemin Zhang

Smart Power Systems and Smart Grids

Also of Interest

5G.
An Introduction to the 5th Generation Mobile Networks
Ulrich Trick, 2020
ISBN 978-3-11-072437-0, e-ISBN 978-3-11-072450-9

Communication and Power Engineering
R. Rajesh, B. Mathivanan, 2017
ISBN 978-3-11-046860-1, e-ISBN 978-3-11-046960-8

Power Systems and Smart Energies
Faouzi Derbel, Nabil Derbel, Olfa Kanoun, 2017
ISBN 978-3-11-044615-9, e-ISBN 978-3-11-044841-2

Digital Electronic Circuits.
Principles and Practices
Shuqin Lou, Chunling Yang, 2019
ISBN 978-3-11-061466-4, e-ISBN 978-3-11-061491-6

Wind Turbine Airfoils and Blades.
Optimization Design Theory
Jin Chen, Quan Wang, Zhenye Sun, 2017
ISBN 978-3-11-034421-9, e-ISBN 978-3-11-034438-7

Qiang Lu, Ying Chen, Xuemin Zhang

Smart Power Systems and Smart Grids

Toward Multi-Objective Optimization in Dispatching

DE GRUYTER

清華大學出版社
TSINGHUA UNIVERSITY PRESS

Authors
Prof. Qiang Lu
Department of Electrical Engineering
Tsinghua University
Haidian District
30 Shuangqing Road
100084 Beijing
P. R. China
luqiang@tsinghua.edu.cn

Ying Chen
Department of Electrical Engineering
Tsinghua University
Haidian District
30 Shuangqing Road
100084 Beijing
P. R. China
chen_ying@mail.tsinghua.edu.cn

Xuemin Zhang
Department of Electrical Engineering
Tsinghua University
Haidian District
30 Shuangqing Road
100084 Beijing
P. R. China
zxm_ts_ee@163.com

ISBN 978-3-11-044784-2
e-ISBN (PDF) 978-3-11-044882-5
e-ISBN (EPUB) 978-3-11-044802-3

Library of Congress Control Number: 2021946972

Bibliographic information published by the Deutsche Nationalbibliothek
The Deutsche Nationalbibliothek lists this publication in the Deutsche Nationalbibliografie; detailed bibliographic data are available on the Internet at http://dnb.dnb.de.

Cover image: metamorworks/iStock/Getty Images
Typesetting: Integra Software Services Pvt. Ltd.
Printing and binding: CPI books GmbH, Leck

www.degruyter.com

Preface

With the developments of communication technology, measurement technology, computer technology, control theory, and applications of new primary and secondary devices (e.g., flexible alternating current transmission system devices, nonlinear optimization subduer (no power system stabilizers), dispersed reactive power compensator (distribution static synchronous compensator), filter, and phaser measurement units) in power systems, people have higher demands and expectations for modern and future power systems. Researchers around the world have proposed a series of concepts such as digital power system [1], smart power system [2–4], intelligent dispatching control center, advanced dispatching automation system, smart grid, and micro-grid to describe the possible forms of future power systems.

It has been a long time since these concepts have been proposed. Nonetheless, some issues still need to be clarified, for example, the relationships among the digital power system, smart power system and smart grid; definitions, principles, and objectives of these systems; and how these systems are implemented. In this book, the smart power system is defined as the initial stage of the digital power system. The concept of a smart power system is one of the subconcepts of the digital power system. We claim that the smart power system is a power system with a multi-index optimality-approximating capability. In other words, the power system with such capability is smart and intelligent. This book presents the relationships between smart power systems and smart grids proposed by former President Obama in terms of concepts, frameworks, principles, configurations, and project practice.

The book comprises nine chapters. Chapter 1 describes the basic concepts, structures, and significance of smart power systems. Meanwhile, it provides a brief introduction to the relevant researches. We hope to make it easier for a reader to comprehend the possible forms of future power systems.

Chapter 2 introduces the basic theory, namely power system hybrid control theory, to accomplish the objectives listed in Chapter 1. It should be noted that it is hard to achieve multi-index optimality-approximating power system operations with existing methods since the power system is a high-dimensional nonlinear system. The authors of this book offer an effective way to solve this problem. The Event-driven methodology is recommended to eliminate the unsatisfied state to achieve multi-index optimality-approximating operations. As a consequence, power system hybrid control theory is the theoretical basis for the smart power system (at least dispatching of the smart power system).

Chapters 3 and 4 introduce the infrastructure and basic platforms of smart power systems. In Chapter 3, the infrastructures of the smart power system are introduced, including digital substations, digital power plants, and digital transmission lines. On the one hand, they provide data sources for global information sharing. On the other hand, they provide means for coordinative control. They are indispensable components for information exchange in smart power systems.

https://doi.org/10.1515/9783110448825-202

Chapter 4 introduces the basic communication platform and data-sharing platform. The former is a physical transmission platform for the exchange of data. The requirements, architecture, and key technologies of the basic communication platform are illustrated in Chapter 4 in detail. A data-sharing platform is a logical transmission platform for the integration of data. It is composed of a universal data access interface, common information model, and data exchange platform. The independence between the data access method and the application is achieved through the universal data access interface during data exchange. The independence between data representation and application is achieved through a common information model. The independence between data processing and data locating is achieved by the data exchange platform. Data processing and data locating share a common data-sharing mechanism, thereby improving the consistency and validity of the data.

Chapters 5–7 describe the key technologies in smart power systems. In Chapter 5, a smart power system operation index system is proposed. It is used to evaluate whether the system is in multi-index optimality-approximating operation states. Key technologies (advanced state estimation algorithm and optimal power flow algorithm based on constraint transformation technology) for event processing and analysis in smart power systems are introduced in Chapter 6. Chapter 7 introduces the visualization techniques of a smart power system, a combination of machine intelligence and the artificial intelligence of system operators.

Chapter 8 presents the smart energy management system (SEMS). In this chapter, the definition, characteristics, fundamental algorithms, and structures of the SEMS are introduced.

Chapter 9 introduces some definitions, contents, and new technologies related to the smart grid which are of common concern to academia and industry at present, including distributed energy technology, large storage and long-life energy storage technology, new technology of grid interaction with users, and operation control of distribution network, especially new technology of network reconstruction (self-healing). The proposal and development of these new technologies enrich the regulation methods of distribution networks and users. It contributes to improving the reliability, economy, and quality of power supply, enhancing the interaction ability of users in the power supply, and developing the smart power system.

In sum, this book is intended to forecast future trends of development of power systems combining theory and practice, show the orientation of technical innovation in the field of dispatching automation of power systems, and offer an effective solution to the construction of smart power systems and smart grids in China.

This book summarizes the author's research on smart power systems and smart grids in recent years under the national 973 Project, the National Natural Science Foundation of China, the State Grid Corporation of China, and the South Grid Corporation of China.

Smart power systems and smart grids are emerging subjects and fields. It is still at a developmental stage. Further research efforts and innovations are required from researchers and practicing engineers worldwide.

Contents

Chapter 1
Introduction

1.1 Review of the concepts of digital power systems

The purpose of this introductory chapter is to review the concepts of digital power systems (DPS).

As defined by Qiang [1] in 2000, a DPS is a digital, visual and real-time description and reappearance of the physical structure, technical characteristics, management, and personal information of a physical power system that is in operation. This definition merely emphasizes a DPS's acknowledgment, understanding, and reflection of a physical power system without addressing the DPS's ability to refine and improve such physical power systems. Therefore, the concept and functionalities of a DPS should involve two aspects:

(1) All necessary data of the system should be acquired in real-time. "Status infor mation" is obtained from the advanced state estimation units in smart energy management systems (smart EMSs or SEMSs). The obtained information can be amicably transferred to the operator and regulator.
(2) The control and operating signals are generated through SEMSs based on the acquired information. The operation is performed automatically to improve the performance of the physical power systems.

Based on the foregoing discussion, the concept of a DPS can be more precisely defined as follows: An operational power system is considered to be a physical power system. A DPS is the digital, visual and real-time description and reappearance of the physical structure, technical characteristics, management, and personal information of a physical power system. The control decisions and recommendations for the physical systems can be obtained based on the acquired data in the physical power systems. The control operations are performed automatically to achieve multi-index optimization and to enhance the capability of management. Physical power systems that involve the abovementioned attributes and functionalities are considered to be DPSs.

According to Qiang [1], a DPS should contain the following features and functionalities:

The physical structure of the power system, the physical characteristics of every component in the power system, the operation mode and operation strategy, and the information of the personnel in the power system.

The measurements and the communication of the real-time state variables in each component or at each node.

https://doi.org/10.1515/9783110448825-001

The settings of all the automated control devices and their function characteristics (including relay protection devices).

The operation status and the "health" conditions of major apparatus in generation plants and substations.

Economic structure and market information.

Special natural conditions (such as thunderstorms, frozen rains, and hurricanes) affect the secured operation of the power system and the real-time evaluation of the impact of such conditions on power system operation.

Information about managers and technicians.

The indices of environmental protection and the monitoring of the commitment and operation of environmental protection equipment.

Coal consumption per kWh for fossil-fuel plants and the emission of SO_x, NO_x, and sniffable particles.

The automated closed-loop control of the physical power system to achieve multi-index optimality-approximating.[1]

The contents of DPSs cover all businesses for all types of power generation and power grid companies, such as real-time dispatching and control, environmental protection, enterprise management, operational management, sales management, intellectual property, and human resource management. The tasks that are assigned to DPSs are therefore comprehensive and complicated. The indices of these optimization-oriented operations will be inevitably abundant. Tremendous amounts of effort and time will be required to realize all indices. Therefore, the best approach is to realize these indices stage by stage. Those stages with the most urgent demand should be realized first. In this book, the tasks and functionalities of DPSs in the narrowest sense (or in the primary stage) are defined from the viewpoint of power system operation and real-time dispatching. These primary stages, which are most urgent, are called "smart power systems (SPSs)."

1.2 Definition of smart power system

1.2.1 Smart power systems and smart wide-area robots

As previously mentioned, for DPSs, the primary stages that are most urgent are called "SPSs." Here, the urgency refers not only to the needs for a nation but also to the demand of the development of the power system itself.

An SPS is precisely defined as follows:

1 This statement is made by us and does not appear in Reference [1].

Definition 1.1 An SPS is a power system with a multi-index optimality-approximating capability.

Why is an SPS defined as such? Consider the objective of an SPS. Which type of operation is preferred, a "better" operation or a "worse" one? The answer is a "better" operation. In addition, the operation characteristic of the power system is expected to be as "good" as possible. From a mathematical perspective or intuitively, such an expectation ("as good as possible") is an "optimization" issue. Now, an indisputable declaration can be made: the issue of an SPS is, in fact, an "optimization" issue.

Is this optimization issue multi-index? Based on demand, power system optimization issues can be classified into at least three major categories: security, quality, and cost-efficiency. Therefore, the optimization issue in a power system is multi-index.

Why, then, do we emphasize in defining an SPS that the system must have "the capability of optimality-approximating operation?" The intelligence of military commanders is demonstrated solely through their organizational and command abilities during military battles (operational behavior) and through the results of such battles. Similarly, the level of intelligence of a human-made system can be assessed only by its performance in the completion of a planned mission (operational behavior). If a pilotless aircraft can select the optimal path, retrieve the maximum extent of useful information from an enemy, and optimally escape from the enemy's following system with minimum energy consumption, the aircraft can be referred to as the "smartest air robot."

It is not difficult to understand that a human-made power system has no intelligence. When SEMS is provided and realization elements that can automatically optimally achieve multiple objectives are installed, the power system may be considered "smart." These types of power systems may be considered smart robots. However, the type of robots mentioned here is not a generalized form. These robots are enormous human-made wide-area robots. Our goal is to provide these robots with intelligence so that they can become smart wide-area robots.

1.2.2 SEMS and smart power systems in China

To our delight, with the support of the State Grid Corporation of China (SGCC), a project involving the construction of an automatic intelligent power system voltage and volt-ampere reactive control system for the three northeastern provinces in China was completed before February 14, 2010, the date of the Chinese Spring Festival. The project was dominated by the Northeastern Grid Corporation as its innovation core organization. Furthermore, the first phase of the SEMS project, which was dominated by the Shanghai Grid Corporation for the Shanghai Power Company, was completed and accepted in November 2007. In early 2009, the second phase of the SEMS project in Shanghai was launched. Domestic and foreign experts visited the Shanghai Power Company to investigate it. SEMSs have gradually become increasingly popular throughout the years. A plan for the construction and later operation of a SEMS in Shenzhen was initiated. The project of the smart Fujian Grid was planned to begin in 2010 with the support of the SGCC. Meanwhile, there was also a plan to begin the project of the smart Hainan Grid with the support of the China Southern Power Grid and the new 863 Plan of the Ministry of Technology. Undoubtedly, China is a pioneer in advocating and exploring DPS and SPS concepts through engineering practice.

How can the "smartness" of a SEMS be demonstrated? "Smartness" is reflected primarily through a new automation system targeting multi-index optimality-approximating operation based on the theoretical foundation of hybrid control theory (HCT). Therefore, SEMSs involve instrumentation systems, communication systems, data storage and adjustment systems, analysis and assessment systems, decision and command systems, command execution systems, and combinations of the aforementioned systems. Because SEMSs have continually been improved and developed in recent years, the projects that have been initiated more recently are more advanced than those initiated earlier. It is expected that the newly developed modules with higher technology and reliability will have the plug-and-play capability and will be able to be updated in existing SEMSs because the structure of these modules is highly discrete and standardized. The existing systems can catch up with the newest technology, and therefore, the technology level of systems can grow accordingly.

As previously discussed, because energy management systems (EMSs) are already well known to people in the power industry and "smartness" has consistently been the goal, adding the prefix "smart" to "EMS" is more reasonable than using "EMS" alone. This case is similar to that in which the authors of this book referred to optimal nonlinear excitation controllers of generators as no-power system stabilizers. Furthermore, the term "PSS" is already well understood by the global power system industry.

Herein, we fully explain and clarify the concept and terminologies of SPSs, smart operational control centers, and SEMSs, which were first proposed by scientists in China approximately ten years ago and have been developed throughout the years. This background is beneficial to build a common language among people in academic communication and interaction. The concept of a "smart grid" is described in greater detail in the subsequent chapters.

1.3 The value of SPS construction

An SPS represents the development direction of modern large-scale power systems. The construction of SPSs offers benefits in the form of improvements in security and power quality and a reduction in energy consumption.

1.3.1 Improvement of disaster prevention capability

"Safety first" is the commonly recognized guiding rule for power system operation in China. The rule is a top priority for electric power operators in ensuring the secure operation of a power system. Generally, three aspects are involved in improving the disaster prevention capability of power systems: precontingency prevention, on-contingency control, and postcontingency remedy.

In a SEMS, a real-time online evaluation of the safety conditions of a power system operation is performed. Based on the evaluation results, suggestions for improving the weaknesses of the system are provided. Furthermore, an early warning could be provided to the operators to indicate potentially unstable conditions. This function is called "pre-contingency prevention." Research results show that power systems have a security region Ωs, which is helpful for security dispatching when the boundary of the security region is calculated in real-time. In SPSs, real-time simulation technology may be utilized for the online analysis of the dynamic safety boundary and the visualization of the real-time condition $X(t)$. This technology may provide a thorough understanding of the operational condition of the system for the system operators. Moreover, when the operating point is close to the boundary, warning signals are generated by the SEMS, and the operation modes are adjusted to avoid the deterioration of the condition. The operation modes can also be adjusted by the operators. The abovementioned techniques maintain a significant safety buffer during the normal operation of the system to improve resisting disturbance.

Although the "precontingency prevention" mechanism can improve grid resiliency in the operation of a system, severe accidents may still create contingencies in power systems, hence the motivation for "on-contingency" control. Under these types of contingency conditions, a SEMS must maintain the stability of the system by using clear methods. For most accidents, theoretical analysis, simulated calculation, and operational experience demonstrate that if a contingency control strategy is performed in real-time, the impact of an accident may be constrained to a small range. As a result, the occurrence of power system catastrophic failure is reduced. Generally, an "on-contingency" strategy aims to address three issues: realizing the existing status, providing the corresponding strategy, and confirming the effect when the "on-contingency" strategy is performed. Based on HCT and technology, three solutions to these issues are proposed: developing a dynamic standard information-sharing platform that shares real-time status, developing a smart decision unit based on online real-time simulation, and developing a smart strategy in real-time.

For certain mega-sized, unpredictable, and cascading disturbances and faults, all of the contingency control methods mentioned above may fail. The operator will thus face a postcontingency remedy, which is to carry out the controlled islanding. Thus, the power system automated control center must possess advanced functions to generate an automatic islanding strategy. Furthermore, the optimal black start portfolio may also be included in the scope of postcontingency remedy plans to address extreme cases.

Finally, it can be concluded that an SPS can provide strong technical support during its operation, thereby greatly enhancing the disaster prevention capability of power systems.

1.3.2 Economic operation indices and power quality improvement

To improve the economic operation and power quality of power systems, optimality control strategies of both economic operation and power quality are developed based on the satisfaction of safety and stability requirements. This issue can be divided into the following two aspects:

(1) Assessment of power system economic operation and power quality

Through the online monitoring and analysis of power system operation states, the overall economic profit and power quality level is accurately and promptly assessed. The system's key information, such as voltage states, grid loss, and system profit (with respect to generation, the grid, consumers, etc.), is provided. Based on the information mentioned above, it can be determined whether the power system operation status meets the corresponding economic operation and power quality requirements. When these requirements are not satisfied, control is performed to improve the operation of the power system.

(2) Obtaining the optimal operating point for both physical power system economic operation and power quality

Traditional methods for addressing this subproblem involve using a simple economic dispatch method to determine the base value of the commitment for each generation. Automatic generation control (AGC) and dynamic safety constraints, however, are not considered. Since the efficiency of generation plants is usually not considered in conventional AGC adjustment, the economic operation level of the grid is reduced. Moreover, the safety, stability, or frequency quality constrained conditions may not be satisfied at the new economic operating point. In SPSs, economic operation and AGC are integrated to form hybrid automatic generation control (HAGC), which improves the performance of the entire physical power system. In addition, hybrid automation voltage control (HAVC) is used for smart HAVC instead of conventional AVC. With this control strategy, the optimal operating point with respect to the economy, safety, and power quality may be obtained. Moreover, HAGC and HAVC can be integrated into SPSs to exert optimal control of reactive and active power flow.

An SPS can generate multiple near-optimal control strategies when resolving certain unsatisfied states. Additionally, the operational effect of these control strategies is evaluated using the testing units described in the following chapters. If the pre-evaluated effect meets the associated requirements, the control strategies will be executed automatically. The pre-evaluated results will be submitted to the operation staff. The operation staff still possesses the highest level of operation authority, even in highly automated smart grids.

The control strategies deployed in the SPS employ AVC, AGC, automation quality control, a static var compensator, and a static synchronous compensator. With the help of the abovementioned controls or components, the system will be operated more economically. Furthermore, a smooth adjustment may be attained, thereby enhancing the economic operating level and power quality of the system.

The control of power generation apparatus is an important aspect of multi-index optimality-approximating optimization problems in SPSs. There is no pure SPS because the ultimate goal of the evolution of current power systems is the development of a DPS. From this point of view, it is unscientific to simply state "separate the power plant and power grid." The phrase should be revised to "separate the operation of the plant and the grid while maintaining integrated planning and integrated control." The authors hope that this revised statement will be accepted by power system operators and decision-makers.

It should also be emphasized here that the smart grids in China are power systems in which the voltage level is equal to or lower than 110 kV because these power systems are commonly referred to as "distribution and consumption grids." Operational strategies involving multiple objectives with co-optimality are applicable to these types of systems. These strategies differ from those designed for power systems operating at a voltage level of 220 kV or above. In power systems operating at 110 kV or below, the strategies focus on the coordination between city-level and county-level dispatch. Fluctuating electricity prices and smart meters are used to motivate the participation of load-side consumers to perform peak shaving. This ap proach is the most effective and economical for solving the problem of peak shaving. Enormous economic profit and social value can be realized by using this solution. In addition, smart grids (including the distribution and consumption grids in suburban areas) will not consider AGC and power plant control problems, as mentioned above. Instead, the issues created by distributed hydro, solar, wind, and energy storage devices and microgrids will be considered. The management of these systems is discussed in detail in the following chapters.

1.4 Global research: state of the art

Since the concept of a DPS was proposed in China, researchers have performed many relevant studies, for example, on integrated energy and communication system architectures (IECSAs [5]), seamless communication architectures for power systems, advanced Pennsylvania–New Jersey–Maryland (PJM) control centers, and the intelligent utility network (IUM) and advanced distribution automation (ADA) center of International Business Machines Corporation (IBM). These projects are introduced in the following sections.

1.4.1 IECSA project

The IECSA project [5] was sponsored by the Electric Power Research Institute of the USA and aimed to make full use of the components in a power system that can be used repeatedly or shared. The project is a follow-up study of the utility communication architecture (UCA) and UCA 2.0.

According to the IESCA project report, the high-level objectives are as follows:

(1) Develop a concrete architecture for power industries and relevant vendors that support the function of self-healing in power systems (this objective mainly refers to distribution grids) and the realization of an embedded communication interface for users.
(2) Provide the results of the project to relevant industry-standard committees and associations and establish a robust and open standard for industrial base facilities.
(3) Introduce the concept of system engineering into the development of system architecture and represent the visual architecture by using standardized industry symbols.
(4) Explore and identify the possibility of the collaborative operation between the power system and individual-owned facilities in terms of equipment sharing.

The research results are as follows:

(1) The functional analysis of power system services

IECSA considers the collection and investigation of major power system services to be key in constructing the architecture. The project, based on broad communication among experts in power system service areas, has collected more than 400 use cases and a detailed illustration of the core functions of these cases.

(2) Information environment of IECSA

IECSA defines an environment as the demand for a set of information commonly used with respect to power systems. The "environment" mentioned by IECSA is the information environment, which reflects information requests presented to the system architecture in user cases. IECSA abstracted 20 cases by considering the de mands of configuration, quality of service, and safety and data management. Com munication among substations, operational control centers, customers, distributed energy resource management centers, and external service providers is considered in these cases. Although a more representative set of cases may be selected, this project has drawn sound conclusions regarding the information environment in a power system. Furthermore, a model of information exchange in a power system has been developed.

(3) IECSA, which is independent of any platform model, includes public services, public information models, and common interfaces

In this context, the models and the interfaces are critical to managing distributed control in power systems and the interoperability among the distributed control systems. Here, "service" refers to the basic component of these architectures, and "public service" refers to the abstraction of functions in these services that are used frequently. A "service" is highly repeatable.

The goals of the project are highly attractive; unfortunately, the research results may focus on specific demands. Although many goals are presented, the methods for achieving those goals are not provided. Even the given execution solutions are suitable only in special cases. The project does not provide overall general strategies.

1.4.2 The seamless communication architecture of power systems

The objective of the International Electrotechnical Commission (IEC) Technical Committee 57 is to construct a seamless data communication architecture to achieve data sharing. The major strategies for completing that objective are to establish a standard for transmission EMSs, such as IEC61970 (WG14), and a standard for substation communication automation, such as IEC61850. Similar modeling approaches based on the abovementioned standards are also used in other areas, for example, IEC61969 for distribution dispatch, IEC62350 for distributed energy resource management, and IEC62349 for hydropower modeling. All these standards are designed for data modeling and data exchange in process control and advanced applications in control centers. These are the fundamental communication standards in SPSs. These standards focus simply on describing the content of the corresponding data models, but methods for using those models are not provided. To achieve data sharing, determining how to use these standards must be further investigated.

1.4.3 The advanced control center in PJM

The regional operator (located in the northeastern USA) of PJM has established a smart grid work group together with the transmission authority. The group has developed a smart grid strategy for using an advanced control center [6] as the core functional unit. The key ideas include constructing backup data and a control center whose location should be different from the original control center; utilizing a service-oriented architecture in designing power system applications; building expandable data models based on the IEC61970 standard; ensuring that the application programs are expandable to support larger-scale models; constructing a public model library for studies on operation, market, and power grid planning; and developing historical and future network models.

This project is consistent with the concept of the communication system of power systems proposed in this book. Completing this project will lead to an efficient and convenient information acquisition-delivery-application chain. However, certain essential functions, namely, the formulation of intelligence (multi-index optimality) decisions and the closed-loop real-time execution of these decisions, are not covered.

1.4.4 Intelligent utility network of IBM

The IUN of IBM is a solution for transforming traditional power systems into smart grids [7]. In the first step, the apparatus essential to generation, transmission, distribution, and consumption processes are monitored in real-time. Then, the acquired information is collected through the network. Finally, the operation of the entire power system can be improved by analyzing the acquired information.

IBM believes that a smart grid is a complete structure with information architecture and a fundamental facility network. Power clients, assets, and operations are monitored in the smart grid.

The management, efficiency, and service of power companies can be improved by obtaining information on demand.

The core of the IBM smart grid is responding to demand; responding to demand is consistent with the company culture of IBM. Specifically, IBM thinks a smart grid consists of five aspects: data acquisition, data transmission, information integration, optimization, and information presentation.

(1) Data acquisition

IBM believes that the data source of a smart grid is broader than data acquisition. Real-time data may be derived from three major sources: power grid operational data, asset state data, and customer instrumentation data. Asset state data can be used to evaluate the state of the apparatus, while customer instrumentation data can be used to enhance monitoring of the consumption behaviors of users and to thereby strengthen management on the demand side and provide data for grid planning.

(2) Data transmission

With respect to data transmission, IBM believes that abundant asset state data and consumer instrumentation data should be collected. These data are numerous and dispersed. Such data are unsuitable for acquisition through the supervisory control and data acquisition (SCADA) system in a power system and should be transmitted by Internet Protocol (IP)-based real-time data transmission methods.

(3) Information integration

Information integration is the strength of IBM. IBM has proposed the integration of business information by using an enterprise system bus to satisfy the requirements of smart grids. Information from various application systems will be integrated into a unique data center. The data models should follow IEC 61970/61850/61968 standards.

(4) Analysis and optimization

Analysis and optimization are the core of a smart grid. IBM has categorized smart grids into four hierarchies. The hierarchical structure can provide deep insight into utilizing the corresponding data.

(5) Information presentation

Information presentation refers to the interface between a smart grid and customers or staff. This function includes acquiring information from multiple sources through gateways, customers' itemizing the presentation of the required information based on individual demands, providing display options of the presented information and distributing the calculated results at the enterprise level. IBM has provided information on the key application system and construction ideas in smart grids. In smart grids, senior-level analysis and optimization focus on three areas: asset aging, grid planning, and grid operation. Furthermore, it is necessary to obtain support from key service systems, such as automatic measurement management, remote asset monitoring, mobile work management, and IP-based monitoring and SCADA.

1.4.5 Advanced distribution automation system

The ADA project is an important part of the IECSA project [8]. The objective of the ADA project is to enhance the reliability of the services of distribution and consumption systems for end-level customers, improve power quality, and increase the operational efficiency of power systems. Real-time data preparation, optimal decision, and operational control over generation and transmission are coordinately used in the control of distribution systems. The main functions of ADA include data acquisition with the consistent testing and adjustment of data; completeness testing of distribution models; the modeling and analysis of routinely and event-driven control systems; prevention and alarming; safety and stability analysis; coordinated voltage/var optimization; fault location, isolation, and service restoration; reconfiguration of distribution and consumption network contingency control strategies; and offline pre-management and restoration planning.

To implement the abovementioned functions, interfaces between different systems or databases, real-time dynamic simulation, optimal load forecasting, and real-time operation are necessary.

Indeed, ADA can be considered to be a solution strategy for integrated distribution automation. The approach may provide advanced functions in the near-real-time distribution automation to improve the distribution operation status, thereby possibly enhancing the "self-healing" capability of the distribution grid.

1.5 Summary

The beginning of this chapter reviews the concept of a DPS proposed by the authors approximately ten years ago. The main deficiency in the previous definition of a DPS (i.e., lack of state optimization when the states are known) is then noted. A DPS should incorporate the two previously mentioned aspects.

"Multi-index" problems are then discussed. A complete DPS should contain multiple types of indices, such as power system operation, power transaction, business strategy, environmental protection, business management, development planning, investment decision, innovation and research and development, intelligence cultivation, human organization, and business culture indices. Each type of index can be divided into several detailed subindices. The scope of the core functions of power systems and the expanded scope will be comprehensive. Therefore, it is best to rank the functions into different phases based on the priorities of demand and execute them sequentially. As well known, the most urgent demand for power systems is to ensure safety and to improve quality and energy efficiency in operation. As a result, a series of issues in physical power systems, such as online real-time communication, dispatch, and control, will be involved. These issues must be addressed as the highest priority and should be solved as soon as possible. Solving these issues is the key task in constructing SPSs in China. From this perspective, an SPS can be considered a special type of DPS or a limited objective DPS.

Based on the foregoing discussion, the conceptual differences between DPSs and SPSs are becoming clear. Moreover, the following platforms or systems should be developed: a sensing and communication system, a standardized data modeling system, an information global sharing platform, a dynamic SCADA system, an advanced state estimation unit and real state global sharing platform, an automatic generation of control strategy and automatic execution unit, and a control strategy testing unit.

Section 1.3 discusses how, if the goal of SPSs is achieved, the revenue gained will be encouraging.

In Section 1.4, relevant research and projects are introduced and discussed.

Chapter 2
Overview of power system hybrid control theory

2.1 Introduction

Several optimization indices have been superficially considered for optimizing conventional power systems. However, the "multi-index optimization" problem has largely been avoided. Instead, single-index optimization models have been developed based on the weighted sum method. The solutions to these single-index optimization problems are taken as the sought-after optimal solution. Nonetheless, for large-scale power systems, it is almost impossible to obtain an analytical solution by using the weighted sum single-index optimization strategy under extra-high-dimensional differential-algebraic equations (DAEs) and an inequality constraint, which are the conditions of using an analytical mathematical approach. Therefore, new theories and methodology must be developed to solve such problems.

The hybrid control theory of power systems (HCTPS) is a methodology and theoretical foundation to guide the development of a smart power system (SPS) or future DPS.

For SPSs, we may consider whether a better solution (with a higher level of intelligence) exists for power system optimization problems than the multi-index **optimality-approximating** approach. The answer is no. The multi-index set is *i*-dimensional, and the HCTPS imposes no limit on the number of *i*. Any index that management and system operators deem necessary can be included. After the multi-index I^i_{MUL} set is formed, a control strategy may be automatically generated by a hybrid control system (HCS) to achieve an automatic effect that approaches optimality for multiple indices. The power system is then in a state where the operation requirements are satisfied with *i*-dimensional indices. For establishing a power system with a higher level of intelligence as described, there must be a complete set of methodologies to act as guidance along with a complete set of theories as a foundation. This is where the HCTPS comes into play.

2.2 Basic concepts

In the subsequent chapters of this book, we frequently refer to some basic concepts in control theory. To adapt to the needs of readers with a variety of academic backgrounds, we present a review of state variables, state vectors, state spaces, state points (SPs), state trajectories, state subspaces, state vector fields, and spatial mappings and transformations. In addition, some new concepts in the HCTPS, such as true state vector spaces, are introduced. The key ideas of the above concepts are given below.

https://doi.org/10.1515/9783110448825-002

2.2.1 State variable

When the initial condition $\boldsymbol{X}(t_0)$ of a dynamic system is determined, the dynamic behavior of the system at any time instant $(t > t_0)$ may be determined by a minimum set of variables, $x_1(t), x_2(t), ..., x_n(t)$. State variables are the minimum set of variables that determine the dynamic system behavior. This definition implies that the state variables are linearly independent, meaning that a state variable $x_i(t)$ cannot be linearly represented by other state variables. That is, if $x_i(t)$ is a state variable, then $k_0 x_i(t)$ (where k_0 is a constant) is not a state variable, and vice versa. However, if $x_i(t)$ is a state variable, $dx_i(t)/dt = \dot{x}_i(x)$ can be another state variable.

2.2.2 State vector

$\boldsymbol{X}(t)$ is called the state vector of a system if every state variable $x_i(t)$ of the dynamic system belongs to the vector $\boldsymbol{X}(t)$. The mathematical formula is as follows:

$$\boldsymbol{X}(t) = [x_1(t), x_2(t). ..., x_n(t)]^T \tag{2.1}$$

where $\boldsymbol{X}(t)$ is a state vector of a dynamic system and $x_1(t), x_2(t). ..., x_n(t)$ are all state variables.

2.2.3 State space

The state space is the n-dimensional real space spanned by the components $x_i(i = 1, 2. .., n)$ of the state vector $\boldsymbol{X}$. By convention, the n-dimensional real space is represented by $\boldsymbol{R}^n$ but often denoted by $\boldsymbol{X} \in \boldsymbol{R}^n$.

As the location of any object in three-dimensional space on Earth can be determined by three real coordinates (x, y, z), the state of a dynamic system at any time instant t_0 is determined by a particular point in the state space. The SP in an n-dimensional state space at time t_0 is determined by n coordinates $x_1^0, x_2^0. ..., x_n^0$. The state space of a power system is denoted by $\Omega(\boldsymbol{X})$.

2.2.4 State trajectory

Every point in the state space (SPs) corresponds to a certain state $\boldsymbol{X}_0$ in the dynamic system. An SP changes as $t \to \infty$. The trajectory that SP follows in the state space $\Omega(\boldsymbol{X})$ as a function of time is called the state trajectory. The dynamic behavior (transition process) of the system in a time interval $[t_0,\ t_0 + \Delta t]$ is precisely represented by the state trajectory of the SPs within the given time interval.

2.2.5 State space and state vector field

Given an n-dimensional function vector

$$F(X) = \begin{vmatrix} f_1(x_1, x_2, \ldots x_n) \\ f_2(x_1, x_2, \ldots x_n) \\ \vdots \\ f_n(x_1, x_2, \ldots x_n) \end{vmatrix} \tag{2.2}$$

each component of $\boldsymbol{F}(\boldsymbol{X})$ in (1.2) is a function of the state variables $(x_1, x_2. \ldots ., x_n)$. Every deterministic point SP in the $\boldsymbol{R}^n$ state space $\Omega(\boldsymbol{X})$ corresponds to a deterministic state X^0, which corresponds to a deterministic vector,

$$\boldsymbol{F}(\boldsymbol{X}^0) = [f_1(\boldsymbol{X}^0), f_2(\boldsymbol{X}^0). \ldots ., f_n(\boldsymbol{X}^0)]^{\mathrm{T}}$$

Therefore, every deterministic X^0 in the state space corresponds to a vector $\boldsymbol{F}(\boldsymbol{X}^0)$ determined by the mapping $\boldsymbol{F}(\boldsymbol{X}^0)$, and $\boldsymbol{F}(\boldsymbol{X}^0)$ is called a vector field in the state space. Consider a special vector field where $f_1 = x_1, f_2 = x_2. \ldots .., f_n = x_n$ in (2.2). The vector field in (1.2) can be rewritten as follows:

$$\boldsymbol{F}(\boldsymbol{X}) = \begin{bmatrix} x_1 \\ x_2 \\ \vdots \\ x_n \end{bmatrix} = [x_1, x_2 \ldots, x_n]^{\mathrm{T}} \tag{2.3}$$

The special vector field in (2.3) is the system state vector $\boldsymbol{X}$. In the aforementioned state space definition, the $\boldsymbol{R}^n$ state space $\Omega(\boldsymbol{X})$ is the space spanned by every component of the vector. Therefore, the state vector field is equivalent to the state space. Thus, we mean the "state vector field" when we refer to the state space and vice versa.

2.2.6 Mapping

A point in space $\boldsymbol{Z}$ in $\boldsymbol{R}^n$ can map into a space $\boldsymbol{Y}$. This process is represented by $\boldsymbol{Z} \to \boldsymbol{Y}$ and is read as "space $\boldsymbol{Z}$ maps into $\boldsymbol{Y}$." Space $\boldsymbol{Z}$ and space $\boldsymbol{Y}$ may have the same dimensions, that is,

$$\boldsymbol{Z} \in \boldsymbol{R}^n, \boldsymbol{Y} \in \boldsymbol{R}^n$$

or different dimensions, that is,

$$\boldsymbol{Z} \in \boldsymbol{R}^n, \boldsymbol{Y} \in \boldsymbol{R}^r$$

Generally, $r \leq n$.

Mapping from space $\boldsymbol{Z}$ to space $\boldsymbol{Y}$ can map one point in space $\boldsymbol{Z}$ to one point in space $\boldsymbol{Y}$ ("one-to-one mapping") or can map one point to multiple points ("one-to-many mapping"). Regardless of the mapping types, a specific mapping relationship must exist. This "relationship" is both an "algorithm" and a "transformation" from an algebraic point of view.

For example, a mapping from space $\boldsymbol{Z}$ into space $\boldsymbol{Y}$ can be expressed as the following z transformation:

$$\boldsymbol{Y} = z(\boldsymbol{Z}) \tag{2.4}$$

where $z(\cdot)$ represents a mapping relationship (an algorithm).

The pair of spaces in the mapping can have distinct properties between them. For example, space $\boldsymbol{Z}$ may be a subspace $\boldsymbol{Z} \in \boldsymbol{R}^n$ of an n-dimensional system state space where each coordinate axis of the space is an axis of state variable $x_i(i = 1, 2 \ldots, n)$, whereas space $\boldsymbol{Y}$ may be a control strategy space where each coordinate axis represents a control strategy.

Moreover, there are two kinds of mappings: one-directional (noninvertible) mapping and two-directional mapping. The subsequent chapters of this book focus on one-directional mapping.

2.2.7 True state vector and true state vector field

In the smart management automation system that will be discussed in later chapters, a dynamic supervisory control and data acquisition (SCADA) system obtains data from a data-sharing platform and passes the state data of the physical power system to the ASEU. The state vector of the physical power system described by the state data output from the ASEU is called the true state vector and is denoted by $\boldsymbol{X}_{TS}$.

2.3 The dichotomy of state space

One key idea of the HCTPS is to divide the state space $\Omega(\boldsymbol{X}(t))$ of the physical power system at time instant t into two dichotomous subspaces: the satisfied state space (denoted by $\Omega_s(\boldsymbol{X}(t))$) and the unsatisfied state space (denoted by $\Omega_{\bar{s}}(\boldsymbol{X}(t))$) (see Figure 2.1). To avoid any ambiguity and to improve the optimality of power system operation, the states that lie in between (i.e., the states that are insufficiently satisfactory) are classified into the unsatisfied state space $\Omega_{\bar{s}}(\boldsymbol{X}(t))$. Therefore, the complete state space at time instant t, $\Omega(\boldsymbol{X}(t))$, is equal to the union of the satisfied state space $\Omega_s(\boldsymbol{X}(t))$ and the unsatisfied state space $\Omega_{\bar{s}}(\boldsymbol{X}(t))$ at time instant t:

$$\Omega(\boldsymbol{X}(t)) \triangleq \Omega_s(\boldsymbol{X}(t)) \cup \Omega_{\bar{s}}(\boldsymbol{X}(t)) \tag{2.5}$$

Since (2.5) holds true for any time instant t, this expression is often shortened in the subsequent discussions to

$$\Omega(\boldsymbol{X}) \triangleq \Omega_s(\boldsymbol{X}) \cup \Omega_{\bar{s}}(\boldsymbol{X})$$

To evaluate whether the operating conditions of the power system at time instant t are satisfied, it is necessary to compare $\boldsymbol{X}(t)$ with an established standard index system. The establishment of a standard operation index system will be discussed in Chapter 5. For now, we return to the problem of the dichotomy of state space.

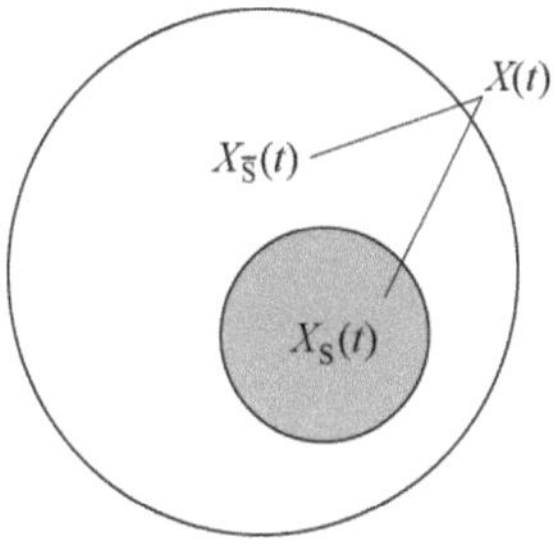

Figure 2.1: The dichotomy of state space.

Definition 2.1 Given an n-dimensional state space $\Omega(\boldsymbol{X})$,

$$\boldsymbol{X} = [x_1, x_2 \ldots, x_n]^{\mathrm{T}} \subset \Omega(\boldsymbol{X})$$

$\boldsymbol{X}$ is the state vector. If none of the state variables $(x_1, x_2 \ldots, x_n)$ of the state vector is unsatisfactory, the state space is called a satisfied state space $\Omega_S(\boldsymbol{X})$; the state $\boldsymbol{X}$ corresponding to any point $P(x_1, x_2 \ldots, x_n)$ in the space is a satisfied state and is denoted by $\boldsymbol{X}_S$.

Definition 2.2 Given an n-dimensional state space $\Omega(\boldsymbol{X})$

$$\boldsymbol{X} = [x_1, x_2 \ldots, x_n]^{\mathrm{T}} \subset \Omega(\boldsymbol{X})$$

$\boldsymbol{X}$ is the state vector. If at least one of the n state variables $(x_1, x_2 \ldots, x_n)$ of the state vector is unsatisfactory, then the state space is called an unsatisfied state space $\Omega_{\bar{s}}(\boldsymbol{X})$; the state $\boldsymbol{X}$ that corresponds to any point $P(x_1, x_2 \ldots, x_n)$ in the space is an unsatisfied state and is denoted by $\boldsymbol{X}_{\bar{S}}$.

Definition 2.3 The unsatisfied state space $\Omega_{\bar{s}}(\boldsymbol{X})$ is called an event space and is denoted by $E(\boldsymbol{X})$:

$$\Omega_{\bar{s}}(\boldsymbol{X}) \overset{def}{\Leftrightarrow} E(\boldsymbol{X}) \tag{2.6}$$

Since the event space is a crucial concept in smart management automation systems, it is necessary to define it precisely.

Definition 2.4 A state subspace of a physical power system at time instant t is called an event space if and only if at least one element $x_i(t)$ of the coordinate of any point $P(x_1, x_2 \ldots, x_n)$ in the space is an unsatisfied state variable $x_i(t) = x_{i\bar{s}}$. The event space at time instant t is denoted by $E(t)$.

Substituting (2.6) into (2.5) yields the following:

$$\Omega(\boldsymbol{X}(t)) = \Omega_S(\boldsymbol{X}(t)) \cup E(\boldsymbol{X}(t)) \tag{2.7}$$

Equation (2.7) shows that the complete state space $\Omega(\boldsymbol{X})$ of a physical power system is equal to the union of the satisfied state space $\Omega_S(\boldsymbol{X})$ and the event space $E(\boldsymbol{X})$. This situation is called the dichotomy of the state space.

2.4 Optimality-approximating state space

Regarding the problem of optimizing power system operations, the authors of this book have always felt that the opinions of practicing administrators and experts deserve particular consideration. Most administration and system operations experts propose three preferred and invariable classes of indices against the objects that they manage or operate (e.g., a physical power system) based on the actual needs of these objects: security and stability indices I_I, power quality indices I_{II}, and the economic operation indices I_{III}; these three classes of indices form a multi-index set:

$$I_{\text{MUL}} = \{I_I, I_{II}, I_{III}\} \tag{2.8}$$

The concept of SPS has been defined as the DPS when only its initial stage and finite objectives are considered. This book includes and discusses only the three classes of indices mentioned above because the three classes of indices are critically important in power system operations.

Definition 2.3 defines the unsatisfied state space $\Omega_{\bar{s}}(\boldsymbol{X})$ as the event space $E(\boldsymbol{X})$. Here, we take a closer look at the event space. Events can be classified into three classes. Accordingly, event spaces are also classified into three categories: $E_I(\boldsymbol{X})$, $E_{II}(\boldsymbol{X})$, and $E_{III}(\boldsymbol{X})$. The union of the three event space classes constitutes the entire event space $E(\boldsymbol{X})$:

$$E(\boldsymbol{X}) = E_I(\boldsymbol{X}) \cup E_{II}(\boldsymbol{X}) \cup E_{III}(\boldsymbol{X}) \tag{2.9}$$

For every event space class (for convenience, we take the class I event space $E_I(\boldsymbol{X})$ as an example), every given point P in the space has a corresponding deterministic r-dimensional event function vector:

$$E_I(\boldsymbol{X}) = \begin{bmatrix} e_{I_1} = e_{I_1}(\boldsymbol{X}) \\ e_{I_2} = e_{I_2}(\boldsymbol{X}) \\ \vdots \\ e_{I_i} = e_{I_i}(\boldsymbol{X}) \\ \vdots \\ e_{I_r} = e_{I_r}(\boldsymbol{X}) \end{bmatrix} \tag{2.10}$$

Each event in (2.10) is a function of the state $\boldsymbol{X}(t)$. To determine whether an event occurs or which event occurs in the power system at time instant t, we need to obtain the state vector of the system at that time instant as accurately as possible:

$$\boldsymbol{X} = [x_1, x_2 \ldots, x_n]^{\mathrm{T}}$$

To this end, essential and regular maintenance of the sensors distributed across the physical power systems and their communication channels is required; more importantly, traditional state estimation technology requires substantial upgrades and thus should be replaced by advanced state estimation (ASE) technology. Detailed illustrations of ASE technology are provided in Chapter 6.

In (2.10), e_{I_i}(i = 1, 2. . ., r) represents class I security and stability events, such as low-frequency oscillation warnings and voltage stability events. e_{I_i}(i = 1, 2. . ., r) is a function of the state vector $\boldsymbol{X} = [x_1, x_2 \ldots, x_n]^{\mathrm{T}}$; therefore, in principle, class I event space $E_I(\boldsymbol{X})$ can be described by a vector composed of r functions or by a vector field. The same holds for the class II and III event spaces.

Given the above basic explanation of the physical and mathematical nature of event spaces, we present a rigorous definition of the optimality-approximating state space.

When the event space at time instant t, $E(\boldsymbol{X}(t))$, is null space, the entire state space of the power system at that time instant is equal to the satisfied state space $\Omega_S(\boldsymbol{X}(t))$, according to (2.7):

$$\Omega(\boldsymbol{X}(t)) = \Omega_S(\boldsymbol{X}(t)), E(\boldsymbol{X}(t)) = 0 \tag{2.11}$$

According to (1.9), $E(\boldsymbol{X}) = 0$ when the three classes of indices $I_{\mathrm{MUL}} = \{I_I, I_{II}, I_{III}\}$ are simultaneously zero. In this case, the complete state space of the power system is equal to the satisfied state space $\Omega_S(\boldsymbol{X}(t))$, according to (1.11). Therefore, it is reasonable to define the multi-index optimality-approximating state space as the state space that satisfies the condition $E(\boldsymbol{X}) = 0$.

Definition 2.5 The state space $\Omega(\boldsymbol{X})$ of the physical power system that satisfies the event space $E(\boldsymbol{X}) = \varnothing$ is called the multi-index optimality-approximating state space, abbreviated to the optimality-approximating state space, and is denoted by $\Omega_{\mathrm{AO}}(\boldsymbol{X})$:

$$\Omega(\boldsymbol{X}) = \Omega_{\mathrm{AO}}(\boldsymbol{X}) \text{ if } E(\boldsymbol{X}) = \varnothing \tag{2.12}$$

In the following, we concentrate on transforming space E into a null space.

2.5 E transform and C transform

When various event spaces $E_Z(Z = I, II, III)$ appear during power system operations, these event spaces are immediately compared on a per-class and per-item basis with the true state vector at time instant t

$$\boldsymbol{X}_{\mathrm{TS}} = [x_{\mathrm{TS}1}, x_{\mathrm{TS}2}. \ldots, x_{\mathrm{TS}n}] \in \Omega(\boldsymbol{X})$$

obtained from the output of the ASEU and with the established power system operation standard index system via the analysis and judgment unit of the smart energy management system (SEMS) to determine whether any event occurred and, if so, which class, subclass, and event type the event belongs to. When an event e $Z_i(Z = I, II, III; i = 1, 2. \ldots, r)$ occurs, it is transformed into a corresponding control command C Z_i with the established DAE equations of the physical power system model and an accurate and effective algorithm. This process is called the E transform. An E transform (for brevity, we are still using the class I event space $E_I(\boldsymbol{X})$ as an example) is represented by

$$C_I(t + \Delta t) = \varepsilon(E_I(\boldsymbol{X}(t))) \; E_I(\boldsymbol{X}(t)) = 0 \tag{2.13}$$

In engineering practice, analysis and judgment necessitate some time to transform the state vector $X(t) \subset E_I(\boldsymbol{X}(t))$ at time instant t into a corresponding control command. This time interval is represented by Δt in (2.13). The research team at Tsinghua University has achieved a $\Delta t < 0.8$ s. Here, we investigate the properties of C_I based on power system practices.

According to (2.10), $E_I(\boldsymbol{X})$ is essentially an r-dimensional vector field, and any element function in the field of $e_{I_i}(\boldsymbol{X})$ may correspond to more than one control command. For example, suppose the analysis unit of SEMS determines event $e_{I_i}(\boldsymbol{X}(\mathrm{t}))$ to be a "low-frequency oscillation warning" event. The decision-making unit may simultaneously issue the following two commands: adjust the setting of automatic generation control (AGC) (to reduce the power transmission appropriately) and adjust the setting of automatic voltage control (AVC) (to improve the receiving-end system voltage quality). Therefore, a single event may trigger more than one control command, such as when events $e_{I_1}, e_{I_2} \ldots e_{I_r}$ occur simultaneously or sequentially, as in (2.10). Generally, C_I in (2.13) is a set of control commands. Using class I events as an example,

$$CI = \{C_{I_1}, C_{I_2} \cdot \ldots, C_{I_r}\} \tag{2.14}$$

Theoretically, it is difficult to eliminate the possibility that class I, II, and III events may occur simultaneously or sequentially. The control command set C Z (***X***) (Z = I, II, III) is the union of the set of control commands corresponding to the different event classes:

$$C_Z = \{C_{I_1}, C_{I_2} \cdot \ldots, C_{I_r}\} \cup \{C_{II_1}, C_{II_2} \cdot \ldots, C_{II_r}\} \cup \{C_{III_1}, C_{III_2} \cdot \ldots, C_{III_r}\} \tag{2.15}$$

Without loss of generality, it is assumed here that $r \neq m \neq s$.

The ε transform in (2.13) is the primary transform in hybrid control, and the resulting C_i is fundamental; however, C_i only indicates the types of tasks to be implemented – not the specific strategies, operating actions, and quantitative information that are required to carry out the tasks.

For example, the decision-making unit of the SEMS may issue a command to reduce the output power of large-scale hydroelectric power stations by 100 MW. However, this unit does not explicitly indicate the adjustments to each generator unit or whether some generator units need to be shut down to meet the requirement. Consequently, a secondary transform is generally required.

We can transform given any control command (using a class I event as an example) $C_i(i = 1, 2, \cdots, j)$ into a corresponding operating order O_i, also called "a set of operations," via the relevant functionality unit of an HCS in the SEMS or distributed autonomous controllers in power plants or substations based on available true state $X_{TS}(t)$ information and reasonable and effective algorithms. This transformation is called a C transform; we have

$$O_i = \mathrm{C}(C_i) \text{ if } C_i = \varnothing; \text{then, } O_i = \varnothing \tag{2.16}$$

where

$$O_i = \{O_{i1}, O_{i2}, \cdots, O_{ij}\} \tag{2.17}$$

We emphasize that the primary objective of O_i is to clear E_i; this objective can be represented by

$$A(O_i) \to E_i = \varnothing \tag{2.18}$$

Equation (2.18) denotes that the result of the action of a set of operations O_i corresponds to transforming the ith event set in class i, E_i, into a null set.

2.6 Geometrical interpretation of the two-level transform

It may be more descriptive and intuitive to understand the ε and C transforms from a geometric point of view. Considering the geometry, the ε as mentioned earlier and

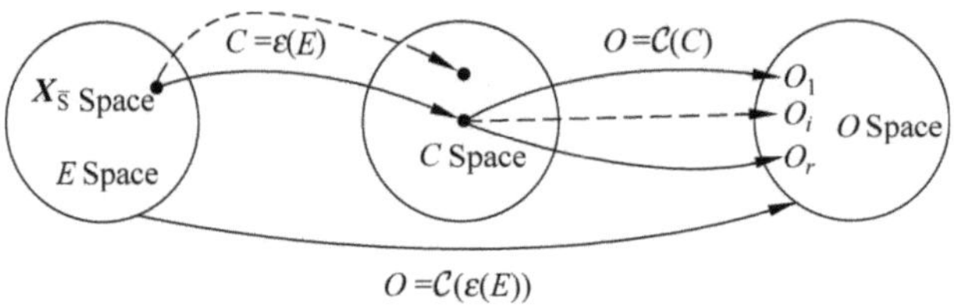

Figure 2.2: Concept graph of the E space and the mapping relationships between the E space and C and O spaces.

c transforms are merely mappings from one space to another. These mapping relationships are illustrated in Figure 2.2.

In Figure 2.2, $C = \varepsilon(E)r$ represents the one-to-one or one-to-many mapping from the event space $\mathrm{E}(\boldsymbol{X})$ to the control command space C; here, ε is the function and $O = C(\boldsymbol{c})$ represents the one-to-many mapping from the **C** space to the operating order space O, where is the function. In addition, $O = C(\varepsilon(E))$ in Figure 2.2 represents a composite mapping from the E space to the O space. Figure 2.2 shows a geometric interpretation of the transforms discussed in Section 2.5.

2.7 Events initiate control; control clears events

The title of this section essentially reflects a philosophical principle that is applicable to power systems and other control systems, especially in multi-index optimization problems intended to solve high-dimensional, nonlinear engineering control system problems.

Definition 2.4 shows that to achieve multi-index optimality-approximating power system operations at time instant t. It is necessary and sufficient to clear the event set E, thereby transforming the event set $E = \{E_1, E_2, E_3\}$ into an empty set. From another perspective, if the physical power system at time instant t has no event, then the state space of the system is equivalent to multi-index optimality-approximating state space. Therefore, we have $\Omega(\boldsymbol{X}(t)) \equiv \Omega(\boldsymbol{X}_{\mathrm{MIAO}}(t))$, and there is no need to initiate hybrid control of the SEMS. Equation (2.8) indicates that the primary ε transform is empty:

$$E_{Zi} = \varnothing \tag{2.19}$$

Furthermore, equation (2.11) shows that the C transform, the secondary transform of the HCS, is also empty under this condition.

$$C_{Zi} = \varnothing \tag{2.20}$$

The operation of the physical power system described above is still monitored by system operators and the traditional EMS (TEMS).

After the HCS in the SEMS obtains the true state vector $\boldsymbol{X}_{TS}(t)$ from the data-sharing platform and the output of the ASEU, conducts an analysis, makes a judgment based on the information, and confirms the occurrence of the event set E_{Zi} of the physical power system at time instant t, the HCS performs an ε transform, $C_{Zi} = \varepsilon(E_{Zi})$, in a timely manner, thereby producing an appropriate C_{Zi} in response to E_{Zi}. Immediately thereafter, the HCS in the SEMS performs the secondary transformation, the C transform $O_{Zi} = (C_{Zi})$, to produce a set of operations $O_{Zi} = \{O_{Zi1}, O_{Zi2}, \cdots, O_{Zij}\}$ to nullify the event set E_{Zi} (to make it an empty set) within a sufficiently short time interval Δt. Then, the physical power system returns to an event-free state, and the HCS in the SEMS has completed a set of control operations without any intervention by system operators. This process may be repeated as mentioned above.

In sum, we can conclude the following:

(1) The spirit of the hybrid control methodology is event-driven; however, the event is the "button" that initiates the HCS.
(2) Once initiated, the HCS in the SEMS continues until the event set becomes an empty set.

We conclude this section by reiterating that events initiate control, and control clears the events.

2.8 Time-driven and event-driven control

Time-driven control is the earliest and basic mechanism in control theory. Concerning power system control, the so-called secondary voltage control proposed by scholars worldwide (and the subsequent tertiary voltage control) is a typical time-driven control. The authors [9–11] correctly realized that this approach would not meet the system requirements if each generator adjusts its output voltage based merely on its own terminal voltage V_{ter}. Therefore, the authors proposed selecting some pilot buses based on certain rules in the controlled power system and focusing on voltage monitoring and control for the selected items. Voltage monitoring and control is the core idea of multistage voltage control. The time interval of adjacent voltage adjustments $\Delta T = T_2 - T_1$ is initially set to 10–15 min and subsequently reduced to 5–10 min or shorter time intervals. However, time-driven control has two main drawbacks: when an adjustment is made at the prescribed time, adjustments actually often prove to be unnecessary. Because most of the voltages at pilot buses are feasible, time-driven adjustments can be redundant and pointless. However, when adjustments are actually needed, the controller may be unable to respond during the period between two consecutive prescribed adjustment points T_i and T_{i+1}. This problem cannot be solved by merely reducing the time interval because the drawbacks are fundamental from the viewpoint of the control theory methodology.

As noted above, the HCT method abandons the time-driven control mindset and adopts the event-driven mindset. Thus, when the controlled physical power system is event-free ($E_{Zi} = \varnothing, i = 1, 2, 3$), the ε transform (as the primary transform) is empty, as discussed in the previous section; consequently, the C transform, $C_{Zi} = \{C_1, C_2, C_3\} = \varnothing$, which is the secondary transform, is also empty, $O_{Zi} = \{O_{i1}, O_{i2}, \cdots, O_{i\omega}\} = \varnothing$.

In a controlled physical power system, the HCS does not perform any operations when no event is present; however, when an event occurs and $E_{Zi} \neq \varnothing$, the HCS must clear the event E_{Zi} and empty the event set by performing a set of timely operations O_{Zi}, after which the physical power system returns to an event-free state or the multi-index optimality-approximating operation state. This process repeats, thereby causing the power system to remain in the multi-index optimality-approximating operation state.

The main feature of event-driven control is event-triggered control. The initiated control actions C_{Zi} and O_{Zi} have only one goal: to clear the event. From this perspective, event-driven control can also be called target-driven control.

2.9 Structure of a power system hybrid control system and smart energy management system

2.9.1 Data source and mathematical model of a power system hybrid control system

The conceptual structure of the HCS is shown in Figure 2.3. From an academic point of view, this figure displays a conceptual structure of an HCS; from a practical engineering point of view, this figure is a conceptual diagram of a SEMS. Figure 2.3 shows that the HCS provides data to the constructed data concentrator through widely installed remote terminal units (RTUs) in the physical power system. Phaser measurement units (PMUs) based on global positioning system technology are rapidly becoming popular in Chinese power systems. Therefore, the data profile $D(T)$ provided by the PMUs has become another source of data input for data concentrators.

Next, we traverse the closed-loop path in Figure 2.3, starting from the data sources (RTU and PMU), and review Section 1.2.2 in the process.

As shown in Figure 2.3, both the top-level decision-making and control and the secondary process and operation levels receive data from the data concentrator. The true state $\boldsymbol{X}_{\mathrm{TS}} \subset \Omega(X)$ is derived from a newly established ASE platform. Then, the ε and transforms are performed by the decision-making and control level (level 1) and the process and operation level (level 2), respectively.

Figure 2.3 shows that the controlled power system at the bottom level is accurately described by DAEs. According to the singular perturbation theory proposed by the French mathematician Henri Poincaré, the algebraic equation $0 = \varnothing(\boldsymbol{x}, \boldsymbol{y})$ can be rewritten as $\varepsilon\dot{\boldsymbol{x}} = \varnothing(\boldsymbol{x}, \boldsymbol{y})$ (where ε is a sufficiently small positive number). Then,

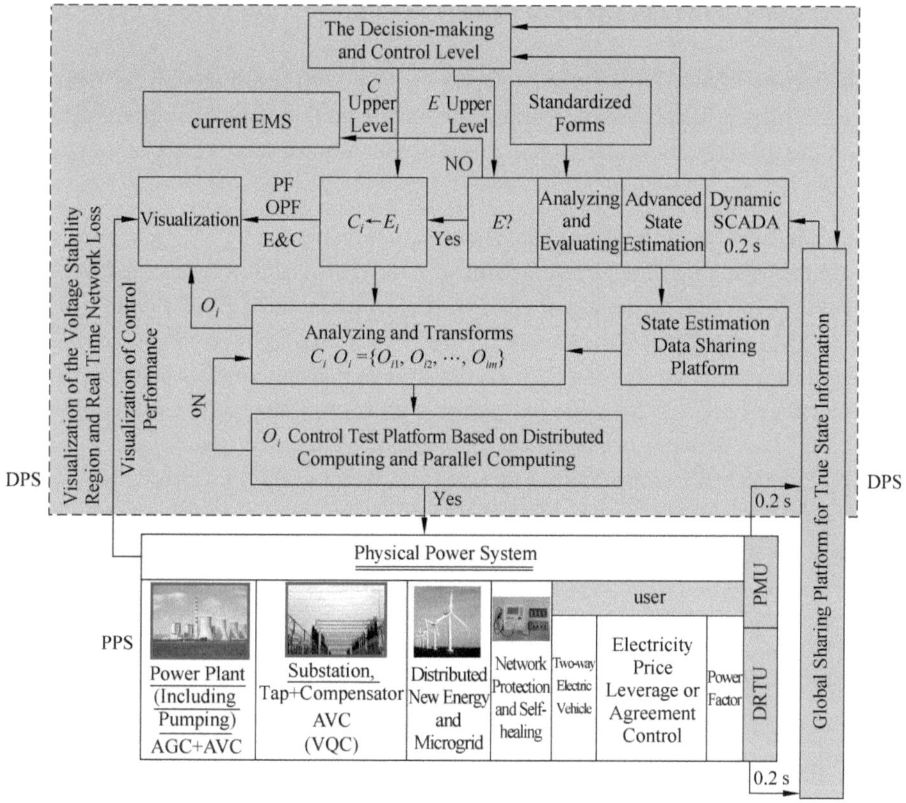

Figure 2.3: The structure diagram of a SEMS based on hybrid control theory.

the behavior of the power system can be accurately described by the following set of high-dimensional, nonlinear, ordinary differential equations:

$$\begin{cases} \dot{x}(t) = f(x(t), y(t)) \\ \varepsilon\dot{x}(t) = \varnothing(x(t), y(t)) \end{cases}. \tag{2.21}$$

where $\boldsymbol{X}(t)$ is the state vector, $\boldsymbol{Y}(t)$ is the output vector, and $\boldsymbol{U}(t)$ is the controlled vector.

Hence, a physical power system is a typical type of dynamic control system. The states of the power system between two adjacent topological structure changes are continuous but not necessarily differentiable (such as the voltage magnitude $V_i(t)$ at bus i).

2.9.2 Analysis of the structure of a SEMS

The definition of an SPS is given in Section 1.2.1. A definition should be clear and concise. Thus, it is necessary to elaborate on the main concepts in the definition and on the primary components that support the definition.

2.9.2.1 Multi-index optimality-approximating operation

As noted above, multi-index refers largely to measures of power system operation security, power quality, and cost-efficiency. Achieving a mathematically rigorous state of multi-index optimality is impractical – and almost impossible. Instead, we use the concept of "multi-index optimality-approximating" in this book because it is both reasonable and possible to operate a power system such that its operating point approaches the multi-index optimality state. The resulting state is called the multi-index optimality-approximating state and denoted by Ω_{MIAO}. Now, we can define the multi-index optimality-approximating capability of a power system. Definition 2.5. The multi-index optimality-approximating control of a power system can automatically maintain state $\boldsymbol{X}(t)$ within the multi-index optimality-approximating state space Ω_{MIAO} during power system operation.

The power system should be able to automatically return to the multi-index optimality-approximating operating conditions when the state of the system leaves Ω_{MIAO} under disturbances by eliminating states that do not satisfy the requirements of the optimality-approximating standard index system (see Chapter 5) through timely and proper control commands and operation actions with the support of global information sharing and ASE technology.

2.9.2.2 Smart energy management system

The term EMS is used worldwide. We are attempting to build a SEMS, which is the cornerstone and primary building block of a smart power system. The term smart energy management refers to the proper response of a SEMS to real-time system state measurements $\boldsymbol{X}_{TS}$. As shown in Figure 2.3, a SEMS needs to determine whether the system is operating within the multi-index optimality-approximating state space by analyzing and evaluating the obtainable power system states. If the system is not operating as such, the SEMS should drive the system back to the optimality-approximating state space Ω_{MIAO} by selecting and performing proper control countermeasures. To this end, a standard index system must be established to determine whether physical power system operating states $\boldsymbol{X}(t)$ at a given time instant lies within the multi-index optimality-approximating state space $\boldsymbol{X}_{MIAO}$. If so, then no control command or operation action needs to be initiated; otherwise, coordinated control applying power system hybrid control theory and technology should be deployed to approach multi-index optimality for the power system, as defined in Definition 2.6. In the definition, a smart power system is defined as a power system with the capability to approach multi-index optimality capability. Therefore, it is not unreasonable to draw an analogy between a SEMS and the perception system and brain of the smart wide-area robot (S-WAR) described in Section 1.2.1. The software blocks (Sections 2.9.2.2– 2.9.2.6) listed in this chapter are the main building blocks of a SEMS.

2.9.2.3 Standardized data modeling and data sharing

The established standardized data modeling and data-sharing platforms provide all the data needed for real-time power system analysis, simulation, and computation in standardized forms in a timely and accurate manner. These standards are the fundamental building blocks that enable global data sharing and the realization of multi-index optimality-approximating control.

2.9.2.4 Dynamic SCADA system

The data-sampling period of a traditional SCADA system is approximately 3–5 s. Such a sampling period is acceptable for traditional EMSs because the sampled data are mainly provided to system operators for system monitoring purposes. However, a sampling period of 3–5 s is too long for SEMSs because the sampled data are directly used for computer-based, closed-loop control immediately after state estimation. The sampling period should be reduced to 0.2–0.3 s; therefore, the sampling rate must be improved by one order of magnitude. We call SCADA with a sampling period of 0.2–0.3 s dynamic SCADA (DSCADA).

2.9.2.5 ASEU

Physical power systems are equipped with many sensors that enable state measurements at virtually every bus. System operators can generate system state data by collecting the aforementioned measurements. This process can be performed by RTU and PMU devices, and the data can be transferred to the SCADA systems in control centers via power-system-specific communication networks.

Most of the data are correct; therefore, their errors are within a standard range. However, for various reasons, the correct data may be mixed with some incorrect data, termed false data (FD). Thus, data filtering is needed even though the FD percentage is small. Hence, we propose the concept of the true state $\boldsymbol{X}_{\mathrm{TS}}(t)$.

Definition 2.6 The state data of a physical power system at a given time instant are regarded as the true state of the system at that time instant if all FD contained in the state data at the given time instant are identified and discarded. The true state is denoted by $X_{\mathrm{TS}}(t)$, $\boldsymbol{X}_{\mathrm{TS}}(t) \in \Omega(\boldsymbol{X})$.

Only the true state $\boldsymbol{X}_{\mathrm{TS}}(t)$ can serve as the source of information and can be used for proper analysis, decision-making generation and operation, and control implementation of the SPS.

As shown in Figure 2.3, the ASEU retrieves data from the DSCADA. The task of the ASEU is to obtain the true state of the system by identifying and discarding FD. The design of the SPS imposes high requirements on the capabilities of an SEU: first, the convergence rate of the calculation must be 100%; second, all FD must be identified; third, the data precision must be greater than the results of all the available algorithms; and fourth, the sources of FD must be identified for maintenance purposes. New ideas and technology are required to create SEUs so that these requirements may be satisfied. Thus, we call a state estimation unit that has such functionalities an ASEU, denoted by ASEU in subsequent chapters.

2.9.2.6 Global sharing of true state information

Since the true state $\boldsymbol{X}_{\mathrm{TS}}(t)=\{\boldsymbol{X}_{\mathrm{TS1}}(t),\boldsymbol{X}_{\mathrm{TS2}}(t),\cdots,\boldsymbol{X}_{\mathrm{TS}n}(t)\}$ obtained through data filtering is required to realize real-time operation and control at different control center levels, a global sharing platform must be constructed to transfer true state information. This platform is the core building block of the smart power system. A question arises: Is the development of the data-sharing platform proposed in Section 2.9.2.2 redundant after building a global sharing platform for true state information? Theoretically, it is. However, such redundancy is probably unavoidable because ASEUs have not yet been widely accepted and utilized; therefore, traditional state estimation units are still prevalent. Often, people's mindsets cause their actions to lag behind new technological developments. This lag is an unfortunate fact that has been repeatedly proven in the history of science and technological development.

2.9.2.7 System to autogenerate centralized control commands and operating orders

SEMSs obtain various real-time operating indices by continuous analysis and computation of true state data and information. Furthermore, SEMSs perform one-to-one comparisons between these indices and the established optimality-approximating standard operating indices. When an instance or set of operating indices does not meet the requirements of the standard operating indices, an instance or a set of events is formulated, and the formulated events are automatically and promptly transformed into control commands, which are further transformed into operating orders by centralized or distributed autonomous control and then executed. These processes are used to improve the operating conditions of the physical power system and clear events. The system to autogenerate centralized control commands and operating orders and that achieves the aforementioned mechanisms and functionalities is the "engine" of the entire SEMS.

2.9.2.8 Control testing platform

All prerelease operating orders should be tested by the control testing platform (CTP) beforehand to determine if they meet the feasibility and effectiveness requirements. When both requirements are met, the orders are executed. Otherwise, a change in the operating orders should be requested. A CTP is built primarily based on accelerated real-time simulation (ARTS) technology. ARTS is needed because the CTP must allot time for control strategy generation.

The SPS in the previously mentioned definitions and descriptions mainly involves the departments and functionalities relevant to the operations, management, and communication of power systems, which include transmission systems and energy management (i.e., national, regional, provincial, city, and county energy management); generation management; substation management; distribution management; distributed generation operation, control, and management; microgrid operation, control, and management; and energy metering, load-side management, and demand response.

As noted above, the definition of an SPS emphasizes the development of a standardized global sharing platform for true state information. All the information needed for decision-making related to analysis, operation, and management needs to be retrieved and processed on the standard data platform. This platform is the foundation of developing smart power systems. Moreover, the definition emphasizes the gradual improvements in the intelligence and real-time scientific decision-making capability of SEMSs. This definition aims to completely replace manual energy management with SEMSs step by step and enable the multi-index optimality of a power system. Finally, the standard data platform fulfills the blueprint for controlling the entire power system (generation, transmission, distribution, supply, and consumption) as an S-WAR.

Next, we answer a question that readers may have been pondering for a while: Why do people call it hybrid control – what does the word "hybrid" mean?

First, as shown in Figure 2.3, the entire control system is a combination of three levels of structures. The top two levels are "digital," and the third level (bottom level) is "physical." This combination of types comprises a "hybrid."

Second, as illustrated in Section 2.8, the control system discussed here is driven by event E_i, which occurs randomly. The time interval ΔT between two subsequent events may range from minutes to hours to days. In short, event E_i is a time-discrete variable; therefore, the composite mapping $O_i = C(\varepsilon(E_i))$ from space E to space O should also be time discrete. However, when a set of operating orders O_i acts on a dynamic power system, the induced power system state variation is continuous. Substituting a set of discrete-time event-induced operating orders O_i into (2.21) leads to the following mathematical model for a physical power system:

$$\begin{cases} \dot{x}(t) = \boldsymbol{f}(\boldsymbol{x}(\boldsymbol{t}), \boldsymbol{y}(\boldsymbol{t}), \boldsymbol{O}_i) \\ \varepsilon\dot{x}(t) = \varnothing(\boldsymbol{x}(\boldsymbol{t}), \boldsymbol{y}(\boldsymbol{t}), \boldsymbol{O}_i) \end{cases}. \tag{2.22}$$

According to (2.22), O_i continuously affects the state $X(t)$ of the controlled power system. Therefore, the entire control system is a typical integration of discrete-time and continuous-time dynamic systems. From this perspective, it is reasonable to call the system a hybrid system.

We discuss one last question in this section. How does the HCS react when multiple events occur? We know that the events can be categorized into three classes, and each class can be divided into multiple subclasses:

$$E=\left\{E_{11}, E_{12}, \cdots, E_{1j}; E_{21}, E_{22}, \cdots, E_{2k}; E_{31}, E_{32}, \cdots, E_{3l}\right\}, j, k, l \in N$$

An event that happens simultaneously in more than one subclass is called a multi-event. Handling multiple events is straightforward. The method assigns different priorities P to the events in each subclass; the events are sorted by P, and the event with the highest P is handled first. Thus, the HCS addresses multiple events very methodically. The power system has very favorable characteristics for multi-index optimality-approximating control. Such characteristics can be considered a type of self-coordination. The so-called self-coordination means that a set of operating orders $O_i=\{O_{i1}, O_{i2}, \cdots, O_{i\omega}\}$ that are effective to clear one event E_{ij} in a particular class might also effectively clear other events (e.g., the reduction in multiple pilot bus voltages in a receiving-end large-scale system; these events belong to the power quality class E_2 event). Objectively, this phenomenon is caused by an insufficient magnetic field intensity at the receiving-end power system and will definitely lead to decreases in the synchronous torque and inertia torque between interconnected systems; thus, this phenomenon is very likely to cause further low-frequency oscillation warnings at the sending-end system relative to the receiving-end system. Because low-frequency oscillations are first-class events (security and stability events), these events are prioritized in the handling process. To this end, the receiving-end voltage level needs to be improved by appropriately increasing the excitation voltages V_f of the relevant generators. This operation is effective at clearing low-frequency oscillations and improving receiving-end system voltage quality. Therefore, during clearing a low-frequency oscillation event (a first-class event), the pilot bus voltage level at the receiving-end system is also improved and qualified. As a result, the second-class event (voltage quality event) may disappear because the power system has the property of self-coordination.

2.10 Summary

As previously discussed in Chapter 1, we should be able to operate and control a large-scale power system as if it were an S-WAR. This is a goal that needs to be fulfilled in the near future.

The robot is a man-made electromechanical system, as is a power system. From this perspective, there are no fundamental differences between robotic systems and power systems, so why do people perceive them as two completely different entities? The answer is because these systems have distinctive appearances. Compared with a typical robot, which occupies little space, a large-scale power system occupies a significantly larger area, and thus, their sizes are not comparable.

Rather than a regular-sized robot, the power system is a special wide-area robot. We are in the process of developing this type of S-WAR. Currently, in China, industrial networks using both fiber-optic and wideband wireless communications have been fully developed for power utility applications and are capable of quickly transferring and storing many data acquired by various sensors (digital and analog) that are implemented and distributed across multiple power plants and substations. The data-sharing platform is used for various investigations and high-speed extraction. Because there are many sensors, it is impossible for every sensor to work simultaneously. Moreover, various types of interference can occur during data transmission, and the data stored in the data-sharing platform inevitably contain FD. For closed-loop controls, it is important to filter the data extracted from the data-sharing platform by the state estimation unit. This is a unique problem in developing the S-WAR for the power system, but this problem can be resolved by applying the ASEUs developed by the authors of this book. When problems pertaining to information technology are resolved, the remaining problems involve improving the system's intelligence. These remaining problems can be resolved by using the multi-index optimality-approximating methodology provided by the power system hybrid control theory elaborated in this chapter.

The structure of the hybrid-control-based SEMS is shown in Figure 2.3, and the SEMS structure is analyzed in Section 2.9. The entire SEMS consists of three levels and six software blocks. The six software blocks include the data-sharing platform (with a data-sharing period of 0.2 s), DSCADA, ASEU, analysis, and event distinction, ε and C transforms, and the control command testing platform.

In conclusion, a new era – in which automated power system management can operate in the same way as an S-WAR – has arrived.

Chapter 3
Smart power system infrastructures

3.1 Introduction

The construction and development of a smart power system (SPS) is a systematic project that covers many aspects, such as power generation, transmission, distribution, and consumption. Therefore, the infrastructures for SPSs are also categorized into several classes, such as digital power plants, digital transmission lines, and digital substations. In this chapter, the definitions of digital substations, digital power plants, and digital transmission lines and the fundamentals for constructing an SPS infrastructure are given. The structure and functionalities of each component introduced above reflect the four basic characteristics of an SPS: standardized modeling, global information sharing, intelligent decision-making, and integrated controls.

The development of digital substations, digital power plants, and digital lines provides useful technical support for constructing smart energy management systems (SEMSs), but these infrastructures are neither sufficient nor necessary conditions for constructing SEMSs. Building a SEMS for a large-scale power system covering a single region or multiple provinces does not require digitalizing all substations, power plants, and transmission lines in the power system region. Specifically, effective bidirectional communications between power plants, substations, transmission/distribution lines, loads, and their control center are key in constructing SEMSs.

3.2 Digital substations

3.2.1 Definition of digital substations

Digital substation technology is the outcome of substation automation (SA) [12]. In traditional substations, secondary devices, such as measuring instruments, signal processing systems, relay protection, automation devices, and telecontrol equipment are isolated from each other. With SA techniques, secondary devices are effectively combined and integrated into multiple smart electronic device groups. Moreover, the monitoring, measuring, controlling, protection, and communication processes for the primary devices and transmission/distribution lines within the substation are automated by using modern IT approaches. SA technology has been widely applied to power grids in China during the last two decades and this technology contributes

https://doi.org/10.1515/9783110448825-003

to maintaining secure and stable power system operations. However, there is room to further develop this technology [13].

1. For historical reasons, the functions developed in conventional SA technology are relatively independent of each other; thus, functions overlap and lack effective coordination. Conventional SA technology also lacks some advanced application modules, such as state estimation, fault detection and analysis, decision-making support, and equipment availability analysis. Furthermore, individual and independent functionalities complicate the SA system design and hamper its efficiency. In severe cases, the fault may spread due to logic confusion in the SA system.
2. Information is not shared between the measurement unit and the control module in current SA technology. This lack of communication may lead to complications in secondary device wiring, communication overloading, difficulties in designing, debugging, and maintenance, and may also lower system reliability.
3. In the current SA system, the nonuniform message formats limit the effective transmission of information as well as the replacement of devices from different manufacturers.

To overcome the deficiencies in the current SA technology mentioned above, some researchers conceived the concept of digital substations in references [14–16]; a digital substation is a type of substation where information is obtained, transmitted, processed, and outputted digitally. Moreover, digital substations have the following features: intelligence of devices, the intelligence of communications, standardized interfaces, and automated operations.

3.2.2 Fundamentals of constructing a digital substation

Both the primary and secondary equipment in a digital substation are intelligent devices, and the information exchanged among these intelligent devices is digitalized using a uniform model. The digital substation is both an important information source and a control terminal for the SPS. The digital substation mainly consists of the following:

(1) Intelligent primary equipment

Conventional technologies are completely replaced by digitalized technologies to effectively improve the performance of data acquisition and driver circuits. This digitalization includes the following: a) applying microprocessors and electrooptical technologies to signal detection and driver circuit operations in primary equipment, b) replacing traditional relays and their corresponding logic circuits with programmable logic controllers, and c) replacing conventional analog electronics and control cables with electrooptic digital signals and optical fibers. In other words, intelligent primary equipment means that the conventional primary equipment is replaced with a digital interface and smart switches. Intelligent primary equipment enables highly accurate and reliable measurement and control and, consequently, guarantees that the information obtained by the secondary equipment is precise.

(2) Internet-capable secondary equipment

Internet-capable secondary equipment is no longer integrated via conventional wirings. An Internet-capable process is based on the exchange of the obtained process data in the secondary device to enable the centralized coordination of the distributed protection and automation function modules. The secondary equipment devices, such as relay protection devices, measurement controllers, remote devices, fault recorders, and reactive power/voltage control devices, are all built on standardized and modulated microprocessors. Communication among devices is based on high-speed Internet instead of repeated input/output interfaces to realize global data and resource sharing. Thus, common function devices are all replaced by logic function modules.

(3) Information sharing platform

An information-sharing platform is constructed based on the digitalization of the substation. This platform provides not only the foundation of the environment of each piece of secondary equipment but also information services for all systems in the control center. The implementation of such a platform enables effective information sharing and provides strong support for application integration.

(4) Automation in operation management systems

There are three primary functions for automatic operation management systems. First, these systems can automatically process data and information from different layers. Second, fault analysis reports and corresponding suggestions can be provided in real-time when a fault occurs in the substation station. Finally, the device maintenance alarm is automatically sent to the substation operators so that the maintenance mode is switched from “periodic maintenance” to “condition-driven maintenance.”

For the four functions mentioned above, the intelligent primary equipment and the information-sharing platform provide technical support to the digital substations, while the coordination of the secondary equipment and the integrated application of the distributed SA functionalities are two important characteristics of the digital substation technology. Specifically, smart switches and smart transformers are key features for the intelligent behavior of primary equipment. Moreover, standardized information interfaces and modeling are essential technologies for the development of Internet-capable secondary equipment, and the automation of substation operations and management is critically important for automating digital substations. However, these goals pose challenges when building a digital substation.

3.2.2.1 Smart switches and smart transformers

Compared with conventional primary equipment, intelligent equipment can effectively receive and respond to commands from an associated control center. Furthermore, self-diagnosis, self-analysis, and self-report procedures can be completed by using intelligent equipment. Intelligent primary equipment mainly refers to two components: smart switches and smart transformers.

(1) Smart switches

Smart switches refer to switches with transducers, controllers, processors, and Internet-capable digital interfaces. In addition to performing a switching function, smart switches also inspect and diagnose digital states. The state inspection functionality includes measuring and monitoring the electric states, magnetic states, temperature, mechanical states, and operations. Furthermore, state estimation and maintenance advice can also be provided. Generally, smart switches have the following features:

(i) high accuracy, high reliability, and low installation costs in the inspection systems;

(ii) the capability to realize intelligent control functions, such as command-triggered switching operations, fixed-phase switching operations, and sequential controls;

(iii) the application of digitalized interfaces to provide switch status information for real-time display and reporting; and

(iv) the improved reliability of the mechanical systems due to the replacement of mechanical energy with electrical energy, the substitution of a mechanical transmission mechanism with motor-driven frequency transformers, and a reduction in the number of mechanical components. Manufacturers beyond China have developed smart circuit breakers that satisfy the International Electrotechnical Commission's (IEC) 61850 standard. These breakers include protection devices and a secondary device at the "process level." This secondary device is used to send the data obtained from the transducers to the data platform of its associated control center and to receive commands from the control center. An important achievement is a digitalization and on-site application of 220 kV smart circuit breakers [17]. In addition to the above functions, the above devices are capable of monitoring current states, such as the integrity of the control circuits, the electrical completeness and continuity of the coils, the kinetic characteristics of movable contacts, the spring status, the mechanical vibration during breaker operations, the number of breaker operations, and the remaining lifetime of the breakers. However, online partial discharging detection techniques are not yet established and are only suitable for field tests.

Currently, smart switch technology is still under development in China. Here, we suggest that 110 kV or above switchyards or substations should adopt a geographic information system (GIS) switching device with intelligent terminals to produce intelligent switches. This device may be the most economical and reliable solution. Secondary equipment is placed at intelligent terminals through the utilization of the GIS control panel to realize self-inspection, interlock control, sequential control, and motor control. In this way, established IT techniques are applied, and conventional switching devices are improved. Using secondary equipment is an economical and effective way to improve the intelligence of switching devices.

The design and development of smart switches have many advantages: the switch status information and commands are transmitted via an optical fiber to save additional wiring on the secondary devices. By enabling conditioned-based maintenance in state inspection systems, the equipment reliability is improved, and the maintenance costs are reduced. The adoptive opening and closing mechanism will lengthen the switch lifetime. Phase-based switching operations can reduce the inrush currents and lower the overvoltage, and the absence of the breaker closing resistor improves the performance of the breakers and enables phase-based switching operations.

(2) Smart transformers

A transformer is a primary piece of equipment used to transfer electrical energy between two or more grids with different voltage levels through electromagnetic induction. Due to the critical role of transformers in power system operations, smart transformers have become a very popular research topic in recent years.

Unlike traditional transformers, smart transformers are typically equipped with transducers, controllers, processors, and Internet-capable digital interfaces. The functions of smart transformers include (in addition to the fundamental functions of conventional transformers) monitoring the operation status and diagnosing abnormal states. The self-monitoring and diagnostic functionalities detect and measure the insulation conditions, thermal states, mechanical status, and so on. Such functionality can be extended so that state assessment and condition-based maintenance can be achieved.

Currently, smart transformers are still under development, without any commercial products available worldwide. In China, digital substations are designed based on conventional transformers. The intelligent module is installed in the transformer terminal box. As a result, the states in the transformers can be monitored and transmitted to a control center via optical fiber, and the commands from the control center can be sent directly to the transformers. An ideal smart transformer is capable of performing a digital inspection of all the important parameters. These measured parameters mainly include the temperature of multiple points on the coils and iron cores, oil level, pressure and temperature, coil voltages and currents, real and reactive power, insulation oil and its decomposition status, partial surge detection, and localization, and cooling system status.

3.2.2.2 Standardized interface and modeling

IEC 61850 *Standards for Communication Networks and Systems in Substations* is a series of standard documents for designating SA and is a part of the IEC Technical Committee 57 (TC 57) reference architecture for electric power systems [18]. IEC 61850 defines the requirements for the information communication behaviors and systems in substations. IEC 61850 also adopts object-oriented modeling technology and a communication-oriented expandable structure as an international standard. IEC 61850 provides the standards for designing a standardized model and interface for a digital substation. We briefly introduce the IEC 61850 standard further.

IEC 61850 has three basic features: (1) realistic interoperability, (2) flexible allocation of functionalities, and (3) high expandability in accordance with advances in communications.

IEC 61850 [19] is a protocol system that fully satisfies the communication requirements for SA systems (SASs), as shown in Figure 3.1. In terms of content, the standards can be divided into four major sections:

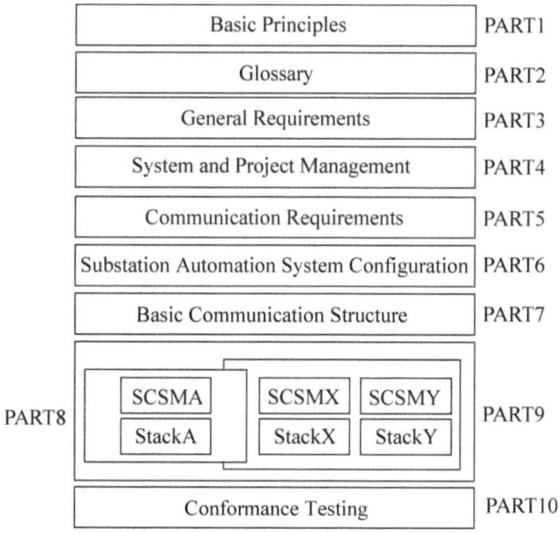

Figure 3.1: IEC 61850 standard.

(1) Systems

This section covers standards from IEC 61850-1 to IEC 61850-5. These five parts offer an introduction and overview of the IEC 61850 standard series. It is clear from these five parts that the IEC 61850 aims to be more suitable for applications in power systems from the point of view of communication technology, project management, quality issues, and system models.

(2) Configurations

In IEC 61850-6, the communication topology structures for SA are given, and the substation configuration description language (SCL) is defined for substations where intelligent primary equipment is included.

(3) Data models, communication services, and mapping

This section covers the key technology in IEC 61850. This section includes seven standards from the IEC 61850-7 series to the IEC 61850-9 series.

IEC 61850-7-3 defines the SAS/supervisory control and data acquisition (SCADA)-based common data class. IEC 61850-7-1/4 defines the logical node (LN) and compatible logical node class (CPLNC) to characterize the basic automation functions. The data object and data attribute, which are generated in the CPLNC instance based on the compatible data class, are used to specify the standard formats for the information exchanged among devices.

IEC 61850-8-x/9-x defines the specific mapping, namely, the communication service mapping, from the abstract communication service interface to the communication networks and protocols. In detail, IEC 61850-8-1 characterizes the mapping for Manufacturing Message Specification. IEC 61850-9-1 describes the mapping for a serial unidirectional multipath point-to-point transmission network, and IEC 61850-9-2 defines the mapping into a process field bus based on IEC 8802-3.

(4) Testing

To guarantee the interoperability and conformance among the multiple devices in the substations, IEC 61850-10 presents standards on the techniques, levels, environment, and equipment requirements for testing the conformance of the substation and devices.

IEC 61850 is the most comprehensive standard series for current substation communication networks and systems. Compared with previous substation communication standards, IEC 61850 has several outstanding technical features: multiple functional layers in the substation, an object-oriented information model, decoupling of functions and communications, the substation configuration languages, and the object-oriented data description.

Currently, IEC is devoted to integrating IEC 61850 and IEC 61970 (standards for communications between substations and control centers) to build IEC 61850+, which will be used for the following communication procedures: the procedure from the process level to the bay level, the procedure from the bay level to the substation level, and the procedure from the substation level to the control center. As the basis for constructing seamless communication systems for power systems, IEC 850+ will be broadly applied to the automation of distribution systems, electric energy metering systems, distributed energy resources, automation systems of solar power, and hydro plants and wind farms. As a result, IEC 850+ will become the most complete and important communication standard. (The IEC will not develop other established communication standards.)

3.2.2.3 Operation and management automation

Another important aspect in digital substation construction is the operation and management automation, meaning that different operations and management scenarios in the substations are accurately simulated by performing comprehensive analysis and data mining on the operational status of the substation and the historical data based on intelligent devices, Internet-capable communication, and standardized models. The utilization of operation and management automation will help improve situational awareness in equipment maintenance and management, increase the accuracy of accident investigations, and improve the technical analysis of equipment to effectively enhance the reliability of substations. The research in this area is still in the initial stage. A brief introduction of the direction of related research is given [20] as follows:

(1) Status inspection and condition-based maintenance

During the maintenance procedure for the primary equipment of the power system, technical development is trending to convert an accident-triggered maintenance mode into a periodic maintenance mode and, further, into a condition-based maintenance mode [20]. Currently, most power system operations are based on periodic maintenance, whose core principle can be described as follows: maintenance must be performed during a fixed period, and the system must recover after maintenance. Practical experiences tell us that the periodic maintenance rule may result in a large waste of funds and resources because much of the maintenance is unnecessary. In contrast, the core concept of condition-based maintenance is described as follows: maintenance is performed when the estimated or monitored system conditions indicate that maintenance is needed and the system must recover after maintenance. Condition-based maintenance will significantly reduce the number of unnecessary maintenance plans and detect hidden deficiencies in the current system in a timely manner to avoid device/system accidents. Thus, condition-based maintenance is an essential approach to improve power system reliability and realize scientific management of the system.

Apparently, condition-based maintenance must be carried out with an accurate device state estimation. All the intelligent devices in the digital substation are capable of sampling and transmitting their operational states and device parameters in a real-time fashion; this capability is necessary to achieve condition-based maintenance. The device states, which include the current operational parameters and conditions, preventive test results, historical operational states of the device, and maintenance records, need to be monitored. Moreover, a tremendous amount of raw data is analyzed and summarized based on a state inspection. Accurate reports on the potential device/system deficiency and faults are provided by applying power system fault detection techniques.

State inspection is an important functionality for digital substations. State inspection can provide data and information, which can be applied for device fault detection, condition-based maintenance, and the optimization of substation and power system operation. Such functionality can lead to significant economic and social benefits.

(2) Condition-based digital risk management system

Condition-based risk management (CBRM) is applied to estimate the device operational states and to investigate the reliability of the power system operation from an industrial application point of view [21]. The CBRM will provide qualitative and quantitative guidance for current and future device maintenance and overhaul schedules. Therefore, CBRM will help the company set short-term and long-term operational maintenance targets and provide decision support to implement condition-based maintenance.

The CBRM system framework includes the following three aspects:

(i) Assessment basis

For CBRM, information and measurements of the target device need to be collected and managed carefully. Then, a comprehensive analysis of the fundamental operational parameters, test results, fault information, and device deficiency is performed by applying statistical principles. Furthermore, the status of the devices and the risk of the grid are assessed according to material aging principles and field operation experiences.

(ii) Assessment procedure

The assessment procedure of the CBRM framework consists of two parts: condition assessment and risk assessment. The CBRM will perform an assessment to quantitatively estimate the current and future device conditions of the primary equipment. The obtained results will be used to predict the outage rate of each power system device, quantify the various risks of each device, and provide different maintenance strategies and plans in consideration of various indices.

(iii) Application of the assessment results

Based on the CBRM assessment results, the device management officers and decision-makers are provided a straightforward understanding of the device conditions from a macro point of view. Appropriate guidance for device management and operational plans, such as condition-based maintenance and technical improvement, may also be given to achieve a reasonable and optimal adjustment. Furthermore, those assessment results may help verify the performance of previous maintenance activities and technical improvements.

In the CBRM framework, the health condition of each device is measured in terms of a health index (HI) and probability of faults (POFs). The risk is in the unit of monetary currency. The corresponding definitions are given as follows:

(1) Health index

The HI is a score ranging from 0 to 10 to quantify the device conditions. This score can be obtained based on the quantification of device information, engineering theory, and field operation experiences.

(2) Probability of faults

The POF refers to the probability that a fault occurs in each individual device and is a numerical representation to reflect the fault possibility trend. The higher the POF is, the more likely the device is to experience a fault. The POF is determined based on the HI, and their relationship is approximated by an exponential function.

(3) Risk

Risk is calculated by the POF, the expected postfault impacts, and the relative importance of the device in the power system. This metric quantitatively represents the degree of risk for the device in a unit of monetary currency.

3.3 Digital power plants

3.3.1 Definition

The construction of digital power plants aims to improve the performance of conventional power plants, particularly thermal power plants.

The key issues to be addressed include inefficient management, high-energy losses, high pollution, unreliable ancillary systems, high outage rates, low automatic protection rates, and redundant staff.

There is no uniform definition for digital power plants. In reference [22], thermal power plants can be regarded as digital thermal plants when plant-level, unit-level, and device-level inspection and management are digitalized. This reference also states that the most important step in constructing digital power plants is to widely promote the utilization of field bus control systems (FCSs). Reference [23] focuses on describing digital power plants from the point of view of management and application. Reference [23] also considers that the goal of digitalization for the thermal power plant is to achieve the measurability of plant management. The final goal of constructing digital power plants is to improve the overall economic efficiency. Scientific plans and designs for all activities in power plants are provided based on advanced information technology and management concepts. Thus, information technology, industrial production, and management are integrated.

Similarly, in reference [24], an analysis of the essential characteristics of a digital power plant is performed, and the detailed features vary among different plants with distinct digitalization depths and widths. A power plant has the characteristics of a digital power plant when each layer in Figure 3.2 is digitalized.

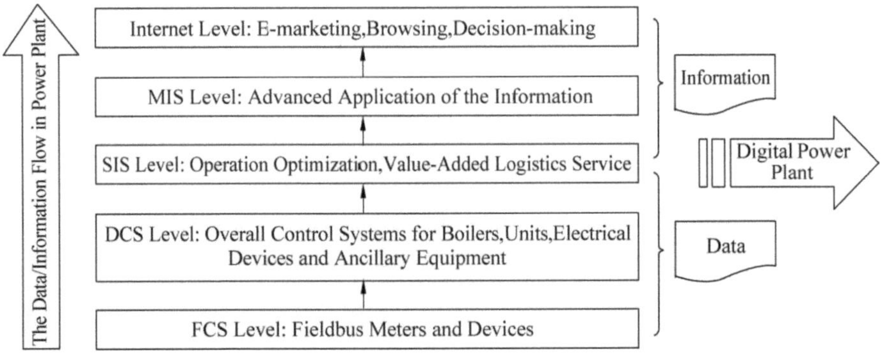

Figure 3.2: Diagram representation of the five layers of data/information flow for a digital power plant.

What are the essential characteristics for a digital power plant that serves as an important regulation module in the realization of the multi-index optimization of SPS operations? What are the objectives of those optimizations? We think that a digital power plant should adopt digitalized modeling and sampling procedures to obtain information of all types and utilize Internet techniques to achieve information exchange and multiplatform resource sharing. These improvements will provide decision-making support for power plant production plans and operations. These improvements will also help increase electricity generation, reduce production costs, and lower environmental impacts and device outage rates. Therefore, compared with conventional power plants, digital power plants are more secure, more efficient, and more environmentally friendly.

3.3.2 Fundamentals of constructing a digital power plant

Currently, many studies on digital power plants adopt a multilevel, multilayer structure. In reference [22], the plant control and operating system are divided into three levels. In reference [24], the plant activities are divided into five levels: Internet level, management and information systems (MIS) level, supervisory information systems (SIS) level, distribution control systems (DCS) level, and FCS level.

Finally, in reference [23], the power plant is divided into two basic platforms (web and database platforms) and three management areas (production control, plant level control, and overall management areas).

In this book, the content for constructing a digital power plant is divided into three levels based on reasons such as the scope, purpose, and data source: direct control, plantwise production control, and production administration and decision-making levels. The structures of these three levels are applicable to thermal power plants and other generation units, such as hydro, nuclear, and wind power resources.

3.3.2.1 Direct control level

The direct control level is directly related to the production equipment and is also responsible for receiving and implementing commands from the plant-level production control level and sending real-time data of units/ancillary devices to the other two levels [25]. Currently, the DCS technique is adopted at the direct control level. The trend of future techniques is the FCS technique.

The DCS is based on utilizing microprocessors and consists primarily of distributed process control, operation management devices, and communication systems. The basic DCS control strategy is to perform distributed controls such that the risks in the systems or the production processes are reduced, and the information is shared for the centralized management. Thus, the DCS allocates the control functions into several process control stations, and the monitoring operations are performed in the monitoring station.

The DCS was introduced in 1970 and has been successfully applied to process control areas for almost 40 years. China started to adopt the DCS in steam turbines and boilers at the end of the 1980s, and DCS significantly improved the automation level of these devices. Since the middle of the 1990s, DCS techniques have been applied to the power system area, with the application of monitoring and control devices in microcomputer protection systems and the incorporation of microcomputers with other automation equipment. The power system information is transmitted to the DCS via hardwire to ensure the reliability, instantaneity, and certainty of the information. As such, the DCS enables the overall controls and monitoring of the boilers, generation units, electrical devices, and ancillary equipment; and this mode of operations is widely used in all new thermal power plants in China. The DCSs in some old thermal power plants, which are still in good condition, have been modified. After 2000, the connection style between the electrical devices and the DCS changed from hardwire to "hardwire+ wireless communication" with the development of Internet-based communication technology. Most information is transmitted via the Internet, while electrical interlock and control information, where high reliability, instantaneity, and certainty are required, are transmitted by hardwires. Such a combination makes full use of the strong data transmission capability of

Internet technologies and guarantees reliable and high-speed data transmission. This method is widely accepted in the industry.

The development target of the DCS is the FCS, which is derived from the DCS. Generally, the nature of the FCS is to utilize intelligent field devices to realize on-site data processing. The primary features for the FCS include the following [24]: (1) complete digitalization (i.e., all the on-site inspection and control devices have their own processors and outputs are in the form of digital serial codes, (2) complete openness (i.e., all the interfaces, external connections, and mutual communications are uniformly standardized to realize the interchangeability of devices from different companies and the generalization of equipment supply sources), and (3) completed distribution (i.e., regular control functions are allocated to intelligent on-site devices). Compared with the DCS, the FCS obtains more information from the field and requires fewer communication wires. On the one hand, the transmission capacity of the communication wires is improved. On the other hand, intelligent on-site devices are used in the FCS; thus, control system functions are independent of the computers or meters in the control center, and the functions can be realized in the field so that the information iteration between fields and the control center is reduced.

3.3.2.2 Plantwise production control level

The plantwise production control level is responsible for monitoring the plant-level production process and serves as the bridge between the production administration and decision-making level with the direct control level [26]. At the direct control level, the commands are broken down and allocated to control the production according to the control information in the production administration and decision-making level and the up-to-date states of each generation unit. At the production administration and decision-making level, the data obtained from the direct control level are organized, analyzed, displayed, and saved to provide the necessary data and information to the production administration and decision-making level.

The functions of the production control level are completed by the SIS at the plant level. The SIS at the plant level manages and supervises production in real-time and determines the globally optimal real-time production schedules. The SIS is based on the DCS. By applying advanced computer network and database techniques, real-time information is shared within the plant. Therefore, SIS is capable of optimizing the plant production procedure so that the reliability and efficiency of the plant are improved. However, in China, there are currently only a few power plants that incorporate the SIS, and the functionalities in use or in the plan include the following:

(1) Production process information inspection, collection, and analysis
The SIS collects real-time information from each production process control system within the plant such that the system can monitor the plant-level production status. Further analysis is performed based on the collected information in the SIS. Then, professional reports and analysis results are provided to management to supervise the whole production procedure.

(2) Optimized allocation among the unit outputs
Optimized allocation of each generation unit output is an important approach to improve the economic efficiency of the power plant. Based on the commands from the automatic generation control systems or other administrations, the SIS can determine the optimal generation unit set and calculate the output of each generation unit to minimize the daily consumption of coal by considering the unit fuel consumption curves and other reliability metrics. Then, the SIS will deliver the computed dispatch results to the control systems of coordination (CSC) of each generation unit and the digital electrohydraulic control devices of the steam turbines to adjust the production plans in real-time. The CSC of each generation unit will respond to the demands from the power grid on the premise of guaranteeing the secure, efficient, and reliable operation of each unit.

(3) Calculation and analysis of the performance in the plant and unit levels
The SIS are also responsible for calculating and analyzing each performance index of the power plant [26]. The plant-level indices include plant operation costs, coal consumption rate for electricity generation, energy production, plant auxiliary power consumption rate, fuel level, oil level, water supply level, steam-water quality, ancillary device steam consumption, and equipment reliability. The unit level indices consist of boiler efficiency, turbine heat loss rate, the efficiencies of the high-pressure, medium-pressure, and low-pressure cylinders, main steam pressure, main steam temperature, reheated steam temperature and pressure, reheater pressure loss, boiler-extracted smoke temperature, oxygen content in the boiler-extracted smoke, the carbon content in the dust, the steam consumption rate of the pump-turbine drivers, the electricity consumption rate of the plant, vacuity in the condensers, supercooled degree of the condensate, supplied water temperature, terminal temperature differences among the heaters, cooling water flow rate in the overheaters and the reheaters, heat rate of the fuels, auxiliary steam consumption rate, unit water supply rate, and steam leakage in the shaft packing.

The economic efficiency indices include the controllable and noncontrollable loss differentials. The controllable loss differentials include main steam pressure and temperature, reheated steam temperature, boiler-extracted smoke temperature, oxygen content in the smoke, the carbon content in the dust, the steam consumption rate of the pump-turbine drivers, electricity consumption rate, vacuity in the condensers, supplied water temperature, terminal temperature differences among the heaters, and the cooling water flow rate in the overheaters and the reheaters. The noncontrollable loss differentials cover reheater pressure losses, heat rate of the fuels, the efficiencies of the high- and medium-pressure cylinders, auxiliary steam consumption rate, unit water supply rate, supercooled degree of the condensate, and steam leakage in the shaft packing. The appropriate and reasonable conversion of the loss differentials mentioned above into the operational costs is helpful to accurately analyze the operational costs of each generation unit.

(4) Equipment state inspection
Equipment state inspection refers to evaluations of the operation status of devices based on real-time operational device information, historical data, and environmental factors. Based on the obtained results, alarms may be automatically triggered for the occurrence of abnormalities, and fault information can be recorded in real-time.

(5) Fault diagnosis and maintenance guidance
Fault diagnosis is based on the information obtained from the equipment status inspection process of the structure and the environmental parameters. This diagnosis aims to achieve fault prediction, fault judgment, and fault analysis. Then, the nature, category, location, degree, and cause of the fault are confirmed. Additionally, corresponding trends and consequences, strategies to control and remove the faults, and approaches to recover the system are generated by the fault diagnosis process. Based on the status information of the primary/ancillary devices obtained from the online supervision systems, SIS is capable of performing online detection of equipment faults. When the fault is detected, detailed coping strategies are provided to guide operators and maintenance staff to adjust the operations and implement maintenance.

3.3.2.3 Production administration and decision-making level

The production administration and decision-making level are responsible for the administration and supervision of the whole plant production and management process [27]. The responsibilities include production procedure management, human resource management, fund management, appropriate utilization of supplies, and cost accounting. Furthermore, at this level, the requirements of operational targets, production goals, and development plans are made based on the company's actual conditions to achieve global optimization. Then, a closed-loop administration is formed [23]. Two primary functions at this level are decision-making and production management.

The production administration and decision-making level, which can be considered the decision-making hub of the digital power plant system, provides decision support for operating and managing the power plant. Therefore, this level should have the following functionalities: (1) commercial operation management, which supports the overall budget, cost management, and computation of operational costs in real-time; (2) closed-loop production and operation management, which is able to precisely supervise the budget-making process, plan an implementation procedure, and provide a performance evaluation and summary; and (3) electricity transaction management, which can provide information on submitted price offerings in the competitive electricity market and some other decision support.

The production administration and operation decision-making level complete its production administration function by applying power plant enterprise resource planning (ERP) systems. The main content of this level is the optimization of the performance indices of the power plant unit and the integration of production plans and strategies. Specifically, based on the operation index information given by the ERP system, the managers in the production administration office can make specific production plans and implement these plans by using the plant-level production management system in the SIS.

3.4 Digital transmission lines

3.4.1 Definition

A transmission line can be considered a digital transmission line when the following conditions are satisfied: (1) all the primary state parameters of the high-voltage transmission lines are measured, transmitted, processed, and exported digitally; (2) a uniform data model and information communication platform are constructed, which enables the real-time monitoring, visualization and analysis of the operational status of the line; and (3) a decision support system is developed, which helps support the line planning, design, operational management, and maintenance scheduling activities.

The transmission line is the most important component in the power grid. High-voltage transmission lines are typically located outdoors or in highly polluted environments. Therefore, dirt retention, lightning, insulator defects, icing issues, swinging, overloading, and overheating may result in the opening act of line breakers, power grid outage, and power interruption. The reliability of the system operation is affected, and the quality of the electricity transmission is reduced. Thus, it is necessary to monitor the real-time operational status of each high-voltage transmission line and provide the obtained information for operators to perform maintenance.

Currently, most system operation centers have no state inspection system specified for transmission lines. A periodic manual patrol for cleaning and maintenance is used for transmission line operational maintenance activity. These activities cost considerable manpower and material resources, yet the effects are unsatisfactory. The construction of digital transmission lines enables the realization of scientific management and decision-making systems for transmission lines. Digital transmission lines help line operators to be aware of the main states and parameters of the transmission lines and their corresponding variation over time. As a result, evidence is provided for scheduling transmission line maintenance plans and cleaning events, which can have great economic benefits. Digital transmission lines also provide technical information about the transmission lines to the system technology engineers for scientific decision-making. The values of the construction of digital transmission lines are reflected in the following three aspects:

(1) The capability of contingent preventive strategies will be significantly improved. The real-time transmission line state inspection system is capable of obtaining more helpful information to prevent accidents and disasters such that the potential post-contingency losses are minimized, and the transmission line security and reliability are greatly improved.
(2) The operation and maintenance costs decrease, and the economic and operation indices improve. Periodic maintenance is changed to condition-based maintenance by real-time state monitoring. Many resources are saved, and the transmission line reliability is improved.
(3) The digitalization of transmission lines provides scientific support that contributes to line planning, design, operation, maintenance, management, and decision-making processes. Thus, the management of transmission lines is improved.

3.4.2 Fundamentals of constructing digital transmission lines

The introduction of digital transmission lines is inevitable. Digital transmission is the result of power system operation and development during a certain period and serves as an important component in SPSs. Generally, there are several primary aspects in the construction of digital transmission lines:

3.4.2.1 Transmission line state inspection

Constructing digital transmission lines requires monitoring their main states, parameters, and operational status in real-time. The further visualization of digitalized line states and parameters is beneficial to the scientific and digitalized management of transmission lines [28]. In fact, the parameters that need to be monitored may vary over different locations. For instance, it is necessary to monitor the insulator pollution level, enhance the inspection of the lightning protection capability in regions with strong thunderstorm activities, and measure the temperatures of the transmission lines with heavy loading during the summer seasons. The transmission line state and parameter inspections are the prerequisites and grounds to realize intelligent management of the lines.

The transmission line states can be categorized into several groups: thermal, mechanical, insulation, and external environmental states. Each group may contain multiple states and parameters. It is not necessary to monitor every state and parameter for each transmission line. However, some important states and parameters must be monitored. Currently, the line states and parameters that need to be monitored include the following:

(1) insulation states, which refer primarily to the lightning protection device status and insulator pollution levels;
(2) thermal states, which include the temperature of the transmission lines, line-fitting devices primarily, and insulators;
(3) mechanical states, which include line-icing conditions, line swing, and windage yaw levels, foundational stresses of towers, tilt degree of towers, and the extent of vibrations; and
(4) external environmental factor inspection, which includes covering detection of weather condition, and surveillance camera systems. For example, manual vandalism events may be inspected by video surveillance.

The insulation states, thermal states, and external environmental factors are the key points of monitoring based on empirical operational experiences because most transmission line contingencies are caused by the worsening of the states or parameters mentioned above.

To enhance the line state monitoring capabilities, an inspection system based on wireless communication and networking technology is applicable for high-voltage transmission lines. The inspection system is integrated with the power plant/SAS to constitute a comprehensive state monitoring system for the whole power grid. Such a system can basically eliminate the need for manual patrols and manual flash-over point detection. Thus, this system provides benefits for precontingent preventive detections and post-contingent reparations. Specifically, the state inspection and visualization systems include several key subsystems:

(1) Lightning location system

The lightning location system is an online system based on Internet technology [29]. This system is used to automatically detect lightning activities over a wide area of the power system footprint in real-time.

As shown in Figure 3.3, The lightning position system includes lightning detection, location calculation, and lightning information services. The lightning position system contains multiple lightning detection stations over the grid, the central station at the control center site, multiple user terminals, and the communication web. High accuracy and efficiency are required in lightning detection and position calculation. For ease of use, lightning event data are processed, presented, and displayed in the information server. The system is capable of receiving, storing, processing, displaying, and sending lightning information in real-time. The system can also present lightning parameters, such as position, time, strength, polarity, and movement of the thunderstorm.

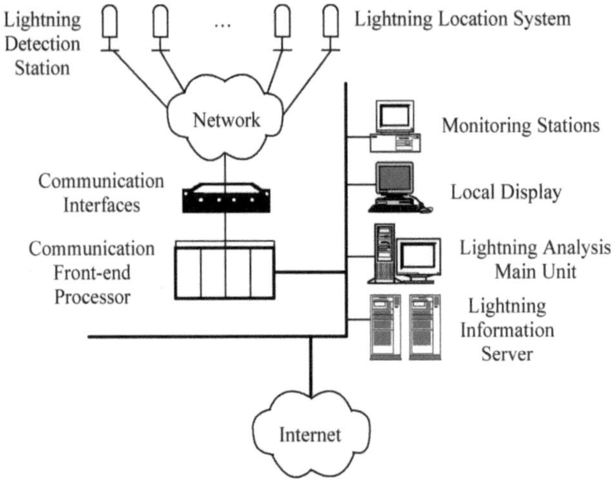

Figure 3.3: Overview of the lightning monitoring system.

The construction of the lightning position system is beneficial for the system operators to identify a lightning-caused fault location as well as the correlations between line-tripping behaviors and lightning activities. The construction of this system is also useful for analyzing and diagnosing transmission line contingencies and transmission line lightning protection design and operation [30].

(2) Online transmission line pollution inspection system

The pollution level on the transmission line surfaces is highly affected by the environmental pollution levels in the area where the line is located [31]. The accumulated pollution on the insulator surface may become wet so that the insulation becomes less effective. Then, an external flashover breakdown may result. Since short circuit faults caused by pollution flashovers may not be cleared successfully, these faults may result in a wide-area power outage. Past experience shows that power system accidents caused by pollution flashovers usually lead to a long outage time and high economic loss. This type of accident is one of the hidden threats to power systems. The historical data show that power system contingencies caused by pollution flashovers are the second most frequent among all the different types of contingencies. Moreover, the losses caused by pollution flashovers are ten times more than those of lightning accidents.

The online transmission line pollution inspection system is shown in Figure 3.4. A set of online inspection devices is installed at a certain distance, and they are used to measure the surface pollution information. Each online inspection device sends a real-time measurement to the power grid control center based on the wireless network, such as the General Packet Radio Service/Global System for Mobile Communications (GPRS/GSM) network. The analysis is performed in the power grid

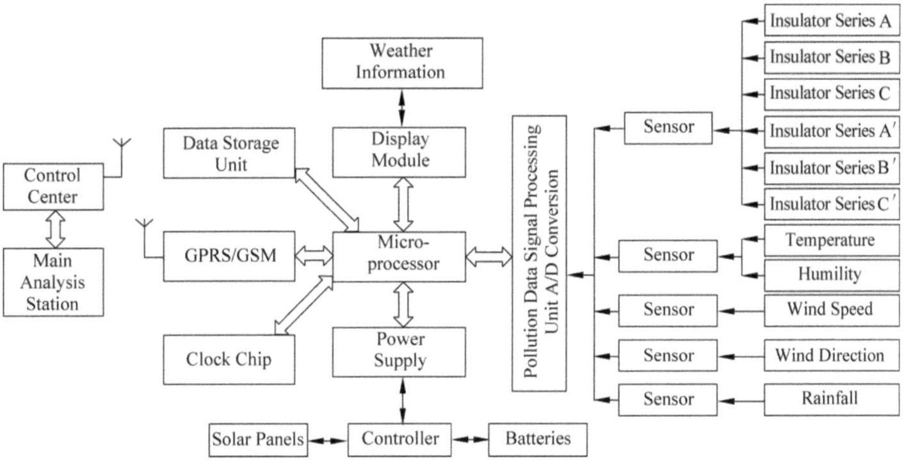

Figure 3.4: Overview of the online transmission line pollution inspection system.

control center by using all the data obtained from the whole grid. Then, the insulator pollution distribution and pollution levels are identified. Other parameters, such as the location temperature, wind speed, and rainfall, can also be measured by this device. These measured parameters will offer support for operators and pollution cleaning staff. Moreover, the pollution accumulation procedure may be reproduced via specific software. In this way, the time when the pollution level reaches a certain threshold is estimated, and the insulator cleaning schedules will correspondingly be determined. The online inspection system has two major parts: distributed transmission line pollution inspection with centralized data management and pollution alarms. These two parts will contribute to improving the scientific management of transmission lines.

The digitalization of pollution monitoring in the construction of the online pollution inspection system has important significance. Digitalization will help to determine the appropriate transmission line cleaning periods, provide scientific guidance for increasing creepage distance and pollution protection, and plot the regional pollution distribution map. Furthermore, suggestions for the design and pollution protection management of transmission lines are also provided.

(3) Transmission line temperature measurement system

The temperature measurement system consists of a series of online temperature measurement devices [32]. Each device is installed on the overhead line and at the line connectors in every local meteorological region. The temperature measurement devices are used to measure the average temperature, node temperature, environmental temperature, and line currents under local weather conditions. Then, the measured data are sent to the main data analysis center via wireless communication devices. The transmission line temperature is useful for determining the line current capacity and allowable line sagging degree; thus, this temperature is an essential parameter for regulating the transmission line power flow capacity. Currently, remote online temperature measurement systems are available at home and abroad.

Data analysis is performed based on the measured environment temperature and SCADA data so that the real-time status of the transmission lines is reported, and technical supports for improving the transmission line capacity and alarm capability is provided. This technique has potentially broad application in the study of line maintenance. The power system coordinators and operators can be aware of the ongoing changes in the transmission line power flows and the margin between the thermal limits and the maximum energy flow rate, thus providing the scientific basis for the dynamic capacity-increase technology of transmission lines. The transmission line operation and maintenance department can perform statistical analysis using real-time loading data and relative historical data to make scheduling plans for transmission line maintenance.

(4) Automatic weather condition measurement system

The issues of insulator pollution, line temperature, insulation conditions, swinging, and icing are directly affected by the local weather and environmental conditions. The measurement of the locational wind speeds, wind direction, environmental temperature, solar insolation, and other weather information can offer scientific decision support for transmission line operation and maintenance.

The weather information measurement system consists of several weather data collection stations. The locations and numbers of stations are determined by the meteorological areas according to line corridors. The location should be close to the local temperature measurement devices. The weather data, such as environmental temperature, locational wind speed, wind direction, rainfall, and solar insolation, are measured by the devices, and the data are then returned to the main station analysis center via the wireless network.

The meteorological administration has already set up a comprehensive network of locational weather data measurement stations, which can provide relatively complete locational weather information. Therefore, the power sector may directly install the automatic weather information receiving terminals provided by the meteorological administration and pay for the maintenance fees to use the available weather data. Thus, there is no need to build a brand-new meteorological information system. Consequently, professional and accurate meteorological data may be provided for a fraction of the cost.

The automatic weather condition measurement system can collect, store, display, remotely transmit, and process meteorological data. This system has three main components: the central supercomputers of the meteorological information network, meteorological data measurement devices, and sensors. The central supercomputers have multiple functions, for example, managing the meteorological information collection network, meteorological information database, and statistical analysis and visualizations. The meteorological data measurement devices are also well equipped to measure, time-tag, store, display, and transmit the local weather data. Until December 2012, the Chinese Meteorological Administration could provide short-term weather forecasts for every 9 km over the nation. This ability is definitely an advantage for us to improve the forecast accuracy of the weather conditions along the main transmission lines.

(5) Video surveillance and automatic alarm systems for transmission lines
Transmission towers and lines are often damaged by natural or outside forces, such as heavy snow, earthquakes, mudslides, and crimes. The damage may also be caused by misbehaviors during nearby construction. Thus, it is necessary to develop video surveillance and automatic alarm systems for transmission lines to reduce the damage caused by natural or outside forces. The systems can also help line operators find the nature, location, and impacts of damage as soon as possible, thereby helping to ensure the reliable and secure operation of power systems.

Advanced digital video compression techniques, low energy consumption methods, GPRS/Code Division Multiple Access (CDMA) wireless communication technologies, and solar power generators can be utilized with video surveillance and automatic alarm systems. On-site video records and environmental information are sent to the control center via the GPRS/CDMA network. This information can also be sent to specific cell phones of line managers via multimedia messages. In this way, the transmission lines and the environment are supervised throughout the whole day. The workload of inspectors is reduced, and the security and stability of the transmission line operation are improved.

The utilization of video surveillance and automatic alarm systems aims to achieve supervision of the following situations: (1) tower base erosion conditions caused by floods; (2) supervision of broken lines or broken strands due to heavy icing or snow cover; (3) unfavorable geological conditions or any other emergency; (4) tilting or collapse of towers; (5) corona and creepage on insulator surfaces and external flashovers; and (6) scissor crossing and sagging overhead transmission lines, erosion of earth wires, swinging situations, and potentially criminal activities.

Based on video surveillance and automatic alarm systems, new line patrol approaches may be found, and a rapid way to find the deficiency or fault of the device is provided. The on-site video surveillance device can also receive commands from the control center at any time to obtain the audio and video records of transmission lines. These are benefits that manual patrol approaches cannot match.

Based on the subsystems described above, we can build a comprehensive transmission line state inspection system consisting of multifunctional probe groups, data storage and transmission terminals, on-site sensor power supplies, data communication networks, data processing systems, control center monitoring platforms, information inquiry systems, and alarm terminals. The on-site multifunctional probe groups will be used to measure the line status and parameters. The latter is converted into digital signals that are then stored. Next, based on the user specified settings, the stored data will be transmitted to the data processing center via the wireless data communication network. Finally, the processed data are sent to servers in the control center. The system is summarized in Figure 3.5.

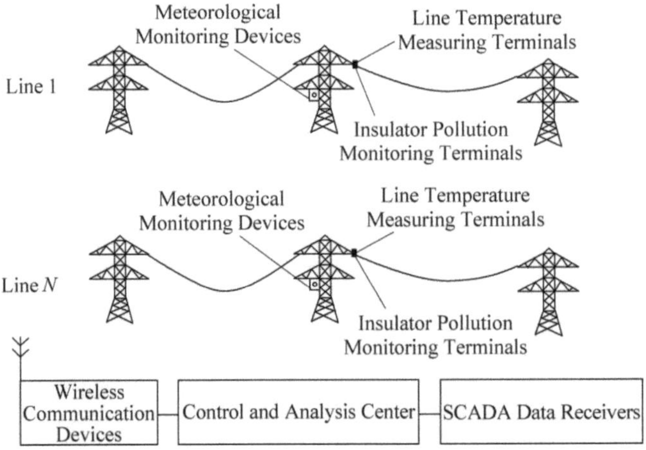

Figure 3.5: Overview of the online transmission line state inspection system.

3.4.2.2 Digitalization of the mobile inspection and management for the transmission lines

Based on the digitalization of the main line states and parameters, personal digital assistant (PDA), GIS, and global position system (GPS) technologies are provided to patrolmen for digitalizing the mobile inspection system and management activities.

(1) Transmission line mobile inspection

Transmission line mobile inspection technology has been applied in some electricity utilities. Mobile inspection technology makes full use of the PDA, equipped with Bluetooth GPS and a GPRS modem. The Bluetooth GPS is used to receive satellite signals. The data are exchanged between the PDA and Bluetooth GPS. The GPRS modem is used to synchronize the data exchange and provide the precise locations of the patrolmen and terminal devices. By utilizing the PDA, Bluetooth GPS, GPRS, and other advanced techniques, there is no need to manually record the observed results during patrol activities. Thus, the patrol activity efficiency is improved substantially.

The transmission line mobile inspection system structure is shown in Figure 3.6, which can also read and display the patrol information on the GIS supervision platform. The towers or facilities with an identified deficiency are highlighted. Then, line

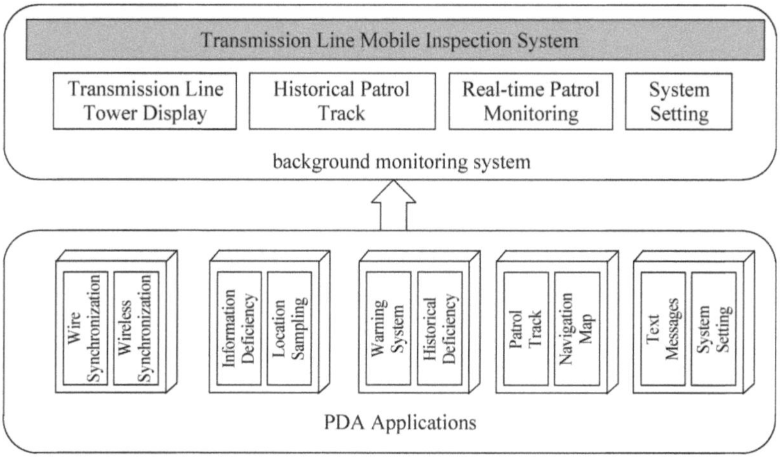

Figure 3.6: Overview of the transmission line mobile inspection system.

operators and maintenance engineers are urged to perform the necessary maintenance work. Apparently, the GIS can also be used by supervisors to verify whether the patrolman is on duty. Real-time inspection of the moving subjects is performed so that the management and patrol quality can be highly improved.

(2) Transmission line management digitalization

Transmission line devices have highly location-dependent characteristics. The application of GIS and GPS technologies will improve the situational awareness of line operators. Furthermore, a comprehensive management system can perform data collection and environmental factor analysis functions to help support decision-making processes for digitalizing daily transmission line management.

First, it is possible to directly present and digitally manage the transmission line information with GIS technology. With the collaboration of MIS and GIS technologies, the line information may be verified on the GIS. Furthermore, the line environments and the condition of the facilities near the lines that are inspected are updated as background graphics in the GIS. In this way, the transmission line management is visualized.

Then, the inspection and reproduction of the transmission line equipment, operational status, and operational environments are realized based on GIS technology. The GIS is integrated with the lightning position system, line pollution inspection system, and management system. For instance, the information of the lightning position system may be displayed on the GIS so that the lightning activity distribution, location, and movement, direction, and strength are clearly shown. Thus, the disaster probability can be evaluated by analyzing the information.

Furthermore, GPS and GIS can also be used for developing 3D transmission line management systems [33]. The data obtained from the background systems are organized, integrated, and transformed. Finally, the data are published on the client-side of the 3D geographic information browsing tools. A schematic diagram of the digital management system is shown in Figure 3.7.

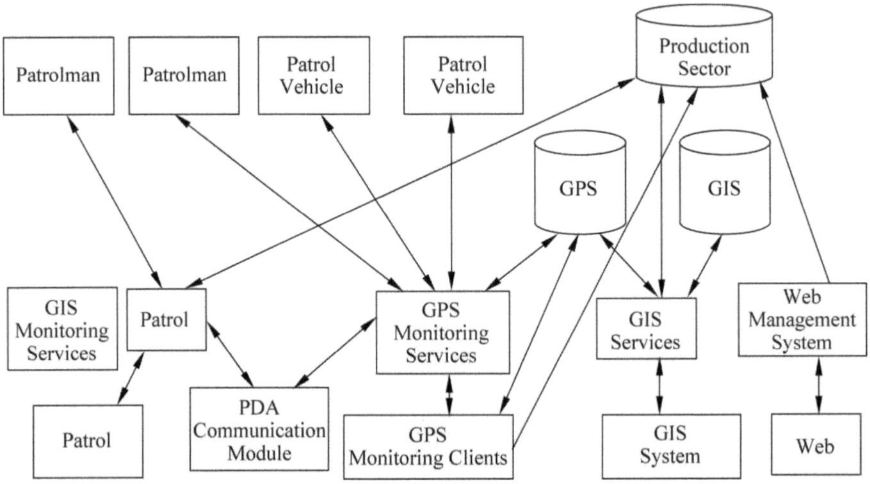

Figure 3.7: Overview of digitalized transmission line management system.

3.4.2.3 Condition-based maintenance for transmission lines

The digitalization of the transmission lines enables the inspection of the transmission line states and the visualization of the operational status; the applications of the PDA, GIS, and GPS technologies enable the digitalization of the intelligent patrol activities and regular management of transmission lines. These achievements will enable fast and accurate measurement for obtaining the technical parameters of transmission lines, finding potential defects, and performing a statistical analysis. Furthermore, these achievements can lay a foundation for condition-based maintenance [34] and intelligent management.

Implementing condition-based maintenance for overhead transmission lines requires two things: First, clear transmission line management content and rules are required. Information management platforms related to device maintenance should be developed. By using the GIS line/substation geographic information management system, the GPS, the GPS-based line patrol system, and other inspection subsystems, the transmission lines are maintained and managed in terms of periodic maintenance, patrol, technical supervision, and security assessment. Second, expert analysis and decision-making systems are indispensable. The precise criteria are proposed by analyzing many raw data. A predictive analysis of device defects is performed, and reasonable maintenance plans are made by considering actual production, operational experiences, and algorithms. As a result, a safe and reliable operation of devices are guaranteed. Furthermore, production costs are reduced. The two aspects shown in the following should be noted during implementation:

(1) In the condition-based maintenance management system, the condition information directly reflects the operational status of every device in the system. It also provides a scientific basis for analysis and maintenance scheduling. The obtained condition information consists of two parts: steady-state information and dynamic information.

(2) In the expert analysis and decision-making system, an expert database should be developed to provide authorized guidance on condition-based maintenance. In addition, the knowledge and experience of experts, national standards, professional rules, and operational experiences are also included in the database. With the development of device and maintenance technology, corresponding correction and amendments are necessary for an expert database.

Based on mastering the general device states, data analysis and condition identification are performed with the help of expert systems. Furthermore, research results related to statistics, mathematics, physics, and high-voltage engineering are used. The above systems should be able to build an intelligent decision-making module for transmission lines to estimate the remaining lifetime of devices, predict accident risk probabilities, and make maintenance strategy plans.

3.5 Summary

In this section, the fundamental facilities for an SPS are introduced. They include digital substations, digital power plants, and digital transmission lines. Unlike traditional power system facilities, new technologies such as uniform modeling, high-performance data collection and communication, comprehensive data analysis, and decision optimization are widely used in digital substations, power plants, and transmission lines. Hence, the capabilities of self-decision-making and coordinative control (or the intelligence of the system) are improved. All the facilities discussed here are the “sense organ” and “limbs” for the SPS. The “brain” and “central” pieces of an SPS are the SEMS, which will be discussed in later chapters.

Chapter 4
Basic platforms of smart power systems

4.1 Introduction

As depicted in Figure 2.1, the basic supporting platform of the smart power system (SPS) is composed of the basic communication platform and the data-sharing platform. Based on these platforms, the data can be shared independently of their sources and applications. Furthermore, the data-sharing platform must be built upon the basic communication platform. This chapter provides a brief introduction to these platforms.

4.2 Basic communication platform

4.2.1 Requirements of the basic communication platform

4.2.1.1 Method of description

For describing the communication needs of the SPS, this book introduces the following six dimensions:

(1) Extensibility

Extensibility measures the system's ability to extend the range of use of the communication channels. In other words, extensibility characterizes the system's ability to support an increasing number of applications, application data flow, and frequency bandwidth of callings.

(2) Availability

Availability is an important indicator of the system's reliability and is usually described by the percentage of services provided by the channels per unit time. Availability is an aggregation of indicators, including the required capacity of the channels, the setting of redundancy, and the resilience of the network (capability of self-healing).

(3) Performance

The performance indicator comprises a collection of indicators, including bandwidth, throughput, accuracy, latency, and delay jitter. Usually, the response latency indicator is the main concern. The so-called response latency is the time that elapses between a sent request and the received response in a communication node. This indicator is not very scientific because it includes the transmission delay on the lines and the time to process the information. Therefore, "message delay" is used instead in

https://doi.org/10.1515/9783110448825-004

the following context. Message delay refers to the time needed to send the data from the data transmission platform via the communication device to the communication channels.

(4) Security

The most basic security indicators include the confidentiality, integrity, and reachability of information. Confidentiality means that information cannot be intercepted by unauthorized persons during communication; integrity means that information is not tampered with during communication and ensures that the information received is consistent with the information sent; and reachability means that authorized data communication can be completed. Security is guaranteed through encryption, intrusion detection, and authorization management.

(5) Management

The management of communication infrastructure typically includes configuration management, security management, and audit management.

(6) Communication modes

Communication usually takes the form of point-to-point, multicast, or broadcast communication.

4.2.1.2 Communication subjects and contents

The construction of the basic communication platform closely links the subsystems, power facilities, and equipment, which are diversely distributed in geography and have different physical functionalities. These subsystems and equipment include the dispatch control centers of the transmission and distribution systems (referred to as the control centers below), substations at all levels, large-sized and medium-sized power plants and small decentralized renewable energy sources, switching equipment and end-users, and a variety of external business systems (such as the company's management information systems and power market system). In addition, the internal communication among some subsystems should also be considered within the scope. This section discusses the performance requirements needed for the communication infrastructure to implement communication among the subsystems.

(1) Communication among control centers

Communication among control centers includes the communication between superior and inferior control centers and the communication among the control centers at the same level. Communication with superior control centers is performed to receive assigned load and generation quotas while submitting part of the operational data to the superior management. Communication with inferior control centers is performed to dispatch load and generation quotas while collecting important operational parameters from inferior dispatch networks. Communication with the same level counterparts is performed to exchange operation parameters.

The communication and exchange of data among control centers are intended to guide the daily production of grids under management, improve the promptness (real-time performance or capability) and accuracy of the analysis, and calculate the operation states (in the system) to ensure the global security and economic operation of the power system. Specifically, the data from other dispatch centers will help improve the accuracy of applications such as state estimation, dynamic power flow, fault analysis, and critical control decision-making. In practice, the above data exchange is realized primarily through texts, spreadsheets, and dispatch phone calls (with recordings). In the future, to meet the development needs of the smart energy management system (EMS), data among all dispatch centers should be exchanged through digital communication in accordance with standardized protocols.

(2) Communication between a control center and power plants

Communication between a control center and power plants is performed to meet both upstream and downstream needs: on the one hand, the control center assigns production tasks and plans as well as control commands to power plants; on the other hand, the power plants send the operational parameters of power generation facilities to the control center.

Communication with power plants is the premise of the output regulation of power plants. In the normal operational state, the control center can send orders to the power plants to adjust the plant output, allocate load in a reasonable manner, and improve the economy of system operation; in a state of emergency, the control center can also directly control the switching of generation units to maintain the stability of the power grid.

The traditional methods of communication with power plants include controlling operating units via telephone (or "remote control" mode) and using telemetry and remote communication mode (or direct exchange of data with the plant monitoring host) to collect various operational parameters within the plant. In SPSs, all the above communication should be achieved through digital monitoring and data exchange. The method used to conduct dispatch should be chosen based on a bilateral agreement.

(3) Communication between a control center and substations

Data exchange is needed between a control center and substations to ensure the secure, stable, and optimality-approximating operation of the power grid. The exchanged information includes upward messages from substations to the control center and downward control commands from the control center to substations. The upward messages include measured values of substations, electric energy consumption within a certain period of time, state information, and sequence of events;

downward control commands include switching operation orders, protection setup modifications, real-time operation parameter callings, and other control commands issued by the control center regarding power quality, load flow, and grid stability.

The above information greatly affects the operation of the power grid; therefore, the security of the communication is very important. In the traditional mode of operation, since decisions are often made by the operation personnel, the decision latency depends on people's response time, which does not require high-speed communication. However, with the development of technology, the decisions of control centers are increasingly made by computer software; thus, the requirements for communication latency have become stricter.

(4) Internal communication platform of a control center

As well known, communication is needed among different application systems within a control center, such as communication among the supervisory control and data acquisition (SCADA) system, human-computer interface, and databases, for both EMS modules and distribution management system (DMS) modules. All exchanged data belong to the category of real-time production dispatch data.

The control center is usually located in a safe and secure physical environment, with a high trust relationship among its application modules. Therefore, the security requirements on the communication platform itself are not too high. However, the exchanged data significantly influence the grid operation; thus, policy-based access control and audit functions should be in place.

As the exchanged information contains critical operation parameters and comprises a very large amount of data, the internal communication platform of the control center should have a short communication latency and high bandwidth. With the development of technology, the ultra-real-time simulation system of the control center will propose higher requirements for the quality of service (QoS) on the communication platform.

(5) Communication between a control center and distributed generations

Communication between a control center and distributed generations (DGs) is similar to that between a control center and power plants. Compared to the impact of the main power plants, the impact of DGs on the operation of the power system is much smaller. Therefore, there is no need to include all DGs in the scope of direct control. Directly controlled DGs have consistent communication requirements with the control center similar to those for the main power plants.

(6) Communication between a control center and users

Communication between a control center and users takes two main forms, namely, meter reading for large customers and direct control of the user loads (under authorization).

(7) Communication between a control center and field equipment

In a few circumstances, a control center can directly control important flexible alternating current transmission system (FACTS) equipment, which in most cases is controlled via a substation. FACTS equipment can effectively improve the distribution of power flow and dynamic characteristics of the system, and a change in the operation states of FACTS equipment may affect the security and stability of the entire system. To avoid misoperations, the control commands issued to FACTS devices by the control center should be accurate and prompt, with high reliability and responsiveness in the corresponding communication process.

(8) Communication inside substations

The contents and performance requirements of such communication have been discussed in the chapter on digital substations.

(9) Communication between substations and field equipment

In addition to the FACTS equipment mentioned previously, the field equipment that communicates with a substation also includes protection devices. This type of communication requires high reliability.

(10) Communication between substations

To meet the needs of stability and reliability at the grid system level, it is necessary for the comprehensive automation systems within the substations to communicate with one another. The data to be exchanged may include data sampling information, phasor measurement results, and real-time control commands. As a complement to the centralized control by the control center, coordination can be decentralized among the substations based on the above data exchange. The core content is that the substations analyze the obtained data in real-time, make control decisions on the spot, and coordinate with other substations during the execution of control commands.

(11) Communication between substations and users

For demand-side management, such as automatic meter reading and real-time pricing, there should be data communication between the end-users and the data acquisition devices of the grid company. The number of data nodes and the amount of data involved are very high and have high requirements for remote configuration and management. Although collecting electricity consumption data from users is less demanding in terms of promptness, the security requirements are high, as the data are closely related to the economic interests of the users and the generation management of the power plants. Accordingly, the communication infrastructure between substations and users should have access controls, auditing, antivirus, integrity checking, encryption, and other functions.

(12) Communication among field devices

For reliable and accurate protection and control, there is a need for high-speed data exchange among intelligent electronic devices (IEDs), including the exchange of measurement data, circuit breaker status, the operation states of devices, and synchronous sampling information. In terms of the operation and control of primary equipment, different aspects of the performance of such communication should be at a high level; otherwise, faults (such as protection failure or malfunctioning and device control failure) can occur, thereby endangering the secure and stable operation of the power system. In previous engineering practice, high-performance communication is often achieved through the use of dedicated communication lines or directly merging related IEDs into a single unit, thereby entailing relatively high equipment costs and inflexibility to extension. With the advances in communication technology, field-bus and network (Ethernet) technology have become mature enough to achieve high-performance communication, providing a more affordable and reliable solution for communication among IEDs.

4.2.1.3 Requirements for performance and functions

The above analysis demonstrates that communication is needed among different parts of SPS with various contents and methods. Table 4.1 summarizes the performance that needs to be achieved in the communication processes mentioned above and their corresponding functional configurations.

4.2.2 Architecture and technology of the basic communication platform

The architecture of the basic communication platform in the SPS is shown in Figure 4.1. The platform is composed of four parts: the basic transmission network, data network platform, clock synchronization system, and security management system.

The basic transmission network covers multiple communication nodes and provides basic channels for the application systems and the data network platform. The transmission media include optical fibers, cables, microwaves, power line communication, satellite communication, and other means. Among these media, optical fiber communication provides high-quality dedicated real-time communication channels and is easy to repair. Thus, it is gradually becoming a mainstream technology.

The data network platform provides communication channels with shared bandwidth and automatic routing, which are suitable for the transmission of large flow, non-real-time application data by means of point-to-point or broadcast communication. Therefore, the data network platform is the main medium of information transmission and sharing in electricity production and management systems.

The clock synchronization system provides a frequency synchronization signal for the synchronous transmission network to maintain link docking among synchronous

Table 4.1: Requirements for performance and functions among different subjects.

	Communication among control centers	**Communication between a control center and power plants**	**Communication between a control center and substations**	**Internal communication platform of a control center**	**Communication between a control center and users**	**Communication between a control center and field equipment**	**Communication between substations**	**Communication between substations and users**	**Communication among field devices**
Extensibility	Needed	Needed	Greatly needed	Needed to some extent	Generally not needed	Generally not needed; increase in a few sites	Depends on sites	Depends on sites	Depends on sites
Availability	99.9%	99.9%	99.9%	99.9%	99.9%	99.999%	99.99%	99.9%	99.999%
Performance	<1 s	<1 s	<1 s	A few $\leq$ 1 ms; mostly $\leq$ 1 s	$\leq$ 1 s	$\leq$ dozens of ms	$\leq$ 1 s	$\leq$ 1 s	$\leq$ 1 s
Security	Integrity Availability Confidentiality	Integrity Availability Confidentiality	Integrity Availability Confidentiality	Integrity Availability Confidentiality	Integrity Availability	Integrity Availability	Integrity Availability	Integrity Availability Confidentiality	Integrity Availability
Management	Security Configuration	Security Audit Configuration	Audit Configuration	Security Audit	Security Audit Configuration	Security Audit Configuration	Security Audit Configuration	Security Audit Configuration	Configuration
Communication modes	Point-to-point, multicast, broadcast	Point-to-point	Point-to-point, multicast	Point-to-point, multicast, broadcast	Point-to-point	Point-to-point, multicast	Point-to-point, multicast	Point-to-point, multicast, broadcast	Point-to-point, multicast, broadcast

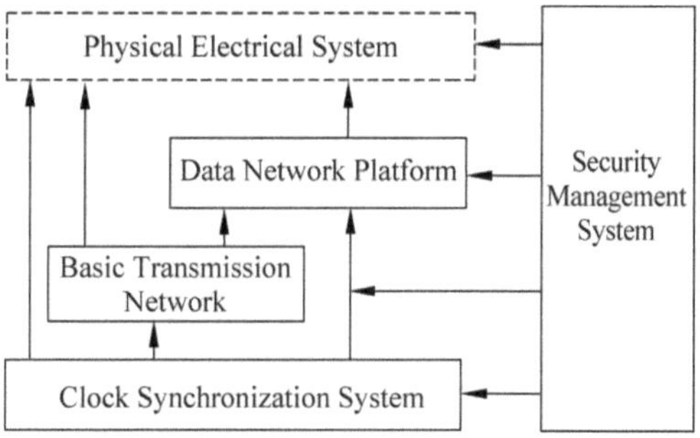

Figure 4.1: Architecture of the basic communication platform.

transmission devices. Due to the increasing demand for real-time capacity in data acquisition and remote control, the clock synchronization system has a prominent role in the SPS. The accuracy of synchronization among different network devices in different regions may directly affect the accuracy of analysis, evaluation, and decision -making of the dispatch system.

Since all the data are digitally collected and processed in the SPS, the security of the communication system guarantees the secure and reliable operation of the entire power system. The security management system ensures the reliability and security of the operation of the communication network, which consistently performs intrusion detection and anomaly detection.

In summary, in the SPS, each part of the basic communication platform has a significant function. The following paragraphs further introduce the time synchronization system, basic transmission network, and data network platform.

4.2.2.1 Time synchronization system

The synchronous transmission network includes a frequency synchronization network and a time synchronization network. Since the frequency synchronization network and time synchronization network share consistent reference sources and precision requirements, it is an inevitable technology trend that the reference sources of two synchronization networks are being merged into one reference source or that two reference sources provide backup to each other. That is, both reference sources provide the time synchronization signal and the frequency synchronization signal simultaneously. As a result, the allocation of resources in the synchronous network is optimized, synchronous network investment is saved, and network maintenance and management are simplified.

(1) Frequency synchronization technology

The present structure of the frequency synchronization network includes four modes: the master-slave synchronization mode, quasi-synchronization mode, mutual synchronization mode, and hybrid synchronization mode.

(i) Master-slave synchronization mode

A master reference clock and several slave clocks are set up in the network, and the frequency of the signals of the slave clocks is controlled by the master reference clock. The master-slave synchronization mode is further divided into a direct master-slave synchronization mode and a hierarchical master-slave synchronization mode. In the hierarchical master-slave synchronous network, the timing signal is transferred hierarchically from the master reference clock to the inferior slave clocks, with each inferior slave clock acquiring the synchronization signal directly from its superior. The advantages of the master-slave synchronization mode are as follows: First, there is no cyclical slide under normal circumstances; second, the requirements for the performance of the slave clocks are low; and third, the cost of establishing the network is low. However, this mode has several disadvantages: an unreliable transmission link may affect the transmission of clock signals, the length of the transmission link is limited due to phase noise superposition, and timing loops may be generated.

(ii) Quasi-synchronization mode

This mode sets up independent high-precision clocks at each node in the network. These clocks have a uniform nominal frequency and tolerance. Each clock operates independently without controlling one other. Although the frequencies of each clock cannot be absolutely identical, the fluctuation error still meets the indicator requirements due to the high-enough frequency precision. The advantages of this mode are its simplicity and flexibility. The drawbacks are the high requirements for clock performance, high cost, and cyclical slides. A data transmission network that adopts this mode of synchronization is called a quasi-synchronous network.

(iii) Mutual synchronization

In the absence of a master reference clock in the network, each clock receives timing signals sent from other nodes and locks its own frequency at the weighted average of the frequencies of the timing signals received. In terms of advantages, this synchronization mode has high reliability and low requirements for the clock performance at each node. In terms of disadvantages, the steady-state frequency is related to network parameters, which are difficult to determine in advance. Therefore, a change in the network parameters will affect the clock frequency and lead to changes in system performance or even system instability.

(iv) Mixed synchronization mode

In China, the mixed synchronization mode is widely used at the current stage. This mode divides the entire network into several synchronous areas. A master reference clock is set up in each area, and slave clocks are set up at each synchronous node. Nodes are synchronized within each area through the master-slave synchronization mode, while areas are synchronized through the quasi-synchronization mode. Through this approach, the number of clock series is reduced, thereby shortening the transmission distance of the timing signals. Furthermore, the mixed synchronization network is much more reliable than a simple master-slave synchronization network due to the reduction in effects related to the transmission distance and external disturbances on the synchronization reference signals. When the master reference clocks in synchronous areas have high precision, the fluctuation error of links among the areas is negligible.

(2) Time synchronization technology

The present mainstream time synchronization technologies include global positioning system (GPS) and related technologies, Internet time synchronization technology, automated computer time service (ACTS), time synchronization technology of shortwave timing and longwave timing, and synchronous digital hierarchy (SDH) network time synchronization technology.

(i) GPS and related technologies

GPS technology is currently a relatively mature time synchronization technology that is used worldwide. However, there are three problems with this technology: First, the GPS system is controlled by the US military. The precision (P) codes are open only to authorized users and the US military. The coarse/acquisition (C/A) codes for civilian use are two orders of magnitude lower than the P codes in terms of time synchronization precision, and their security is not guaranteed. Second, GPS signals are wirelessly transmitted and are susceptible to external interference. Third, the timing signals of GPS receivers are output via nonstandard interfaces. Common network devices (such as switches) have no such dedicated interface. Similar to GPS technology, other timing and positioning systems include GLONASS (GLObal NAvigation Satellite System) of the former Soviet Union, the BeiDou system of China, and the Galileo program of the European Union. Users may choose a system according to their specific situations. It is estimated that the BeiDou system of China will cover the globe and may be used to replace the GPS system of the USA by the end of 2014.

(ii) Internet time synchronization technology

Time synchronization on the Internet is achieved through the Network Time Protocol (NTP) and the Precision Time Protocol (PTP). The standard NTP employs standard Request for Comments (RFC) 1350 standards, while the simplified NTP (SNTP) adopts the RFC 1769 standards. The time resolution of the NTP is 200 ps, which provides 1–50 ms time calibration precision, but experiments have shown that this technology achieves a calibration precision of only a few hundred milliseconds or even on the order of seconds at an intercontinental scale. As prescribed by the Institute of Electrical and Electronics Engineers (IEEE) 1588 standards, the PTP is used for sub-microsecond synchronization of clocks in sensors, actuators, and other terminal devices in standard Ethernet or other distributed bus systems adopting multicast technology.

Compared with the NTP, the PTP improves the synchronization precision and convergence speed, thereby enabling high-precision synchronous control based on the industrial Ethernet. Unfortunately, the PTP has not yet been widely used.

(iii) ACTS technology

ACTS technology requires relatively simple equipment: just phone lines, analog modems, common personal computers, and simple client software. Due to the feedback technology provided, the transmission delay in phone lines can be partially offset. Currently, this technology is mainly used to calibrate the timing of personal computers. Calibration of the clocks of other local devices requires further development of interface hardware and corresponding software. ACTS technology is not real-time. The National Institute of Metrology and the Shaanxi Astronomical Observatory provide this kind of timing service.

(iv) Shortwave and longwave timing time synchronization technology

The technology of using radio to broadcast timing signals has the advantages of wide coverage, relatively simple receiving and sending equipment, affordability, and so on. In contrast to Internet timing technology, it can be used to calibrate local clocks in real-time. Currently, in China, only the Shaanxi Astronomical Observatory provides shortwave timing signals. Internationally, the Loran-C (long-range navigation) system is used in longwave timing. The domestic signal stations are located in the coastal area and are mainly for military use and navigation, not civilian use.

(v) SDH network time synchronization technology

This technology embeds the time-coding signals that are synchronized with a cesium clock into the free bytes of the communication network while keeping the message structure consistent with the international standard protocol. Therefore, as long as the information in the free bytes is not blocked, the long-distance transmission of synchronous time-coding signals can be achieved. There are three methods for transmitting standard time based on the SDH network: the unidirectional method, two-way method, and common-view method.

(3) Construction strategy of the clock synchronization system

Currently, the clock synchronization parts of application systems are separately developed in the systems. The use of clock sources is not uniform, and the devices to be synchronized are incompatible with the synchronization interfaces. As a result, time setting is not accurate, application obstacles may occur in the systems, and calibrating time in the whole network is difficult.

For security and reliability of synchronization of entire networks in a variety of complicated situations, it is necessary to construct a clock synchronization network independent of the business network to support clock synchronization of the entire power system.

As described earlier, the development direction of the clock synchronization system is the merging of the time synchronization and frequency synchronization systems. In other words, it is necessary to set up three systems: the wireless synchronization network, the terrestrial synchronization network, and the local clock source system. These three systems mutually calibrate, thereby providing a unique, precise, and synchronized timing signal for the whole grid and thus guaranteeing a uniform time setting for the entire network.

The wireless synchronization system adopts satellite synchronization technology and currently uses GPS satellite timing signals as the clock source. With the establishment and improvement of China's "BeiDou" satellite synchronization system, this system will become the clock synchronization source of the wireless clock system of the SPS, thereby providing one path of wireless timing signals to the ground nodes.

The terrestrial synchronization system employs a cesium clock and high-grade time servers to provide ground frequency and time synchronization signals for the whole network. The transmission link uses a synchronous or asynchronous data transmission link network. One path of ground timing signals is provided for each ground node.

Each ground station is equipped with a set of time servers with a local frequency for calibration. The three paths of clock signals are put into the calibrator to calibrate the timing signals by the majority principle, thereby excluding the degraded synchronization signals and achieving the purpose of reliable and accurate timing of the entire network.

4.2.2.2 Basic transmission network

The smooth operation of the SPS relies heavily on the communication system and requires that the communication system be built on a solid physical basic network platform. The platform is supposed to have high reliability, with qualities such as resistance to electromagnetic interference and voltage flicker. In addition, the platform should be compatible with a variety of services, such as voice, data, and video communication.

Optical fiber is almost always the best choice, whether it is evaluated from a technical point of view or in terms of cost performance. Additionally, the power system has a unique advantage in the development of optical communication. For instance, the existing channel resources of cables and transmission lines of different voltage levels can be well utilized without the acquisition of new land, thereby avoiding complex social issues and shortening the construction period. Additionally, compared with other options, investment in optical fibers offers cost savings.

The construction of optical fiber networks should be divided into multiple levels, namely (in descending order), the regional power grid, provincial grid, municipal grid, county grid, substations, and feeders. In terms of construction, priority should be given to the first three levels with unified planning. Network construction below the municipal level can be conducted in accordance with the actual situation, following relevant regulations.

It is suggested that optical fiber connections having 48 or more cores be used between regional and provincial grids. Moreover, monocyclic or polycyclic connection methods can be considered in topology design to maximize system reliability. The connections between provincial and municipal networks can combine both ring and star topologies. For critical sites, two or more access lines must be provided. Sites with star connections are usually not sites with heavy data traffic, but other forms of backup communication mechanisms should still be present. Grids below the county level can be laid out following the principles used for municipal grids. In areas where conditions permit, optical fibers should be installed on feeders, and various terminal units should be installed on poles.

To further improve the reliability of the basic communication network, other supplementary forms of communication may also be introduced in necessary areas, such as the following:

(1) Power line carrier (PLC) technology

Using the phase conductor of overhead lines as the transmission medium, a PLC transfers analog or digital voice, telegraph, remote motion control, and remote protection information through a power line. This technology boasts the benefits of high channel reliability, low investment costs, immediate effectiveness, and synchronization with the construction of the power grid.

(2) Microwave communications technology

This technology offers the benefits of large communication capacity, low investment costs (approximately one-fifth of the investment cost of cable), fast construction speed, and resilience to natural disasters. This technology can serve as a backup and supplement a trunk optical fiber network to restore communication when the trunk network is hit by a natural disaster. This technology can also be used in areas and situations that are unsuitable for optical fiber (such as remote mountain areas and islands).

(3) Satellite communications technology

Based on the development of terrestrial microwave communication, global satellite communication can be achieved using relay stations in space. This technology has the benefits of a long communication distance, large coverage area, and high communication flexibility. Because the cost of satellite communication does not depend on the communication distance, satellite communication is extremely cost-effective for outlying cities, rural areas, and areas with poor transportation.

(4) Wireless communication technology

General packet radio service and code division multiple access technologies are wireless communication technologies.

4.2.2.3 Data network platform

The data network platform provides a bearer platform of data services for production systems. This platform meets the needs of the production application system, the data-sharing platform, the informationalization of business management, and personal multimedia applications. Compared with a dedicated line network, the data network has the following advantages:

(1) Bandwidth sharing, which improves transmission efficiency
(2) Intelligent business routing, which increases channel invulnerability
(3) Smart management, which involves diversifying the means of monitoring

Currently, the mainstream technologies used in the data network include asynchronous transfer mode (ATM), Internet protocol (IP) over SDH (packet over synchronous optical networking and synchronous digital hierarchy (POS)), Gigabit Ethernet (GE), and Resilient Packet Ring (RPR). The basic principles, advantages, and disadvantages of each technique are discussed further.

(i) ATM technology

ATM technology is a connection-oriented communication technology that employs a fixed-length information unit as the basic unit of data transmission. A virtual channel is established before data transfer occurs, and the subsequent data transfer is accomplished through this channel. The transfer and forwarding of information units are performed by ATM switches. The ATM switches first analyze the header information in the unit and then, based on the content of the header, forward the unit to the next ATM switch or the final destination. Thus, ATM switches can achieve high-speed transfer and forwarding of information units.

By using the information unit as the basic transmission unit and connecting through virtual channels, an ATM network possesses a good flow control mechanism and QoS while suffering little signal transmission delay. This technology is therefore suitable for supporting and dispatching critical services in the network.

The disadvantages of ATM lie in its protocol complexity, extra expense, and costly devices; it also requires a special operation to accomplish the conversion between information units and data packets to support IP applications. In practice, ATM devices face problems associated with a low port rate (typically 155 Mbps but, in some cases, 622 Mbps), unaffordable prices, unstable performance, and suspended technology development.

(ii) POS technology

Currently, there are two methods for achieving POS technology: one is defined by the Internet Engineering Task Force IP/Point-to-Point Protocol/High-Level Data Link Control/SDH (IETF IP/PPP/HDLC/SDH) framework, and the other is defined in the International Telecommunication Union Telecommunication Standardization Sector (ITU-T) X.85/Y.1321 IP/Link Access Procedure – SDH (LAPS)/SDH framework structure.

POS technology can encapsulate an IP packet of any length into a fixed-length SDH frame. This technology utilizes the existing SDH network to transmit the data and solves the problem of long-distance data transmission in the absence of sufficient optical fiber resources, thereby saving the original investment. In addition, POS technology takes advantage of the resilience and manageability of the existing SDH network to protect communication channels. When a link fails, the POS port can be switched to the protection channel in less than 50 ms. POS technology employs a logical point-to-point connection in the physical SDH ring network, thereby eliminating multistage connections.

The disadvantages of POS are as follows: POS encapsulates variable-length IP packets into fixed-length SDH frames at the sending end and restores these frames at the receiving end, and the decomposition and restoration of packages are complex and costly. Additionally, the congestion control of POS is poor, and POS does not provide quality assurance to end-to-end services; thus, POS is not suitable for multi-priority parallel services.

(iii) Gigabit Ethernet

GE is compatible with fast Ethernet. More than 80% of the world's network nodes are Ethernet nodes. The implementation of GE has a directness, fastness, and gigabit transmission speed. GE redefines the medium access control layer and introduces "carrier extension" to access conflict detection, thereby compensating for the shortcomings of the original Ethernet (such as unreliable topology, as well as with respect to multimedia applications and QoS). Currently, GE technology is not only a switching technology but also a bearer technology that has been widely used in metropolitan area networks (MANs). GE has the benefits of a simple technical route and low-cost equipment. In terms of the drawbacks of GE, it directly uses a bare optical fiber connection with a transport layer, and consequently, it is difficult to achieve optical quality, performance monitoring, and protection; fiber investment is wasted because every two service access points require a pair of optical fibers, and the transmission distance is limited such that the long-distance optical interface can generally transmit only from 50 to 100 km; and in support of key services, there is no perfect solution that would guarantee QoS and security. In addition, the establishment of a GE network relies primarily on two-layer and three-layer routing, and the self-healing and resilience algorithm is slower than other dedicated algorithms.

(iv) RPR technology

RPR is a new packet-based transmission technology that combines the advantages of Ethernet and SDH. RPR combines the efficient use of IP routing technology and bandwidth, rich service integration capabilities, and the high-bandwidth and self-healing capabilities of optical fiber loops to better meet the needs of multiple services of MANs. RPR usually adopts a double ring structure, which is a ring topology composed of two reverse-directed fibers (one clockwise and the other counterclockwise). One node can reach another from both directions on the ring, so RPR is bidirectionally available. Each thread of fiber can be used to transmit data and control signals in the same direction. The use of spatial multiplexing technology further increases the efficiency of bandwidth usage on the ring.

RPR technology has the technical characteristics of spatial multiplexing, ring self-healing protection, automatic topology recognition, multilevel QoS services, a bandwidth fairness mechanism, a congestion control mechanism, and independent physical layer media.

The disadvantage of RPR technology is that its standards were only recently completed. At least temporarily, the standards are supported by only a limited number of mainstream manufacturers; in addition, there are no cross-ring standards, so RPR information in a single ring cannot be passed across rings. As a result, it is difficult to meet the actual needs of a dispatch network (one ring cannot solely meet the needs of services; tangent rings, intersected rings, rings with chains, and other complex network topologies should be available). Therefore, RPR is less capable than SDH in terms of establishing a network.

4.3 Data-sharing platform

4.3.1 The needs of the data-sharing platform

Although the data exchange infrastructure in the existing power system has been achieved, to some extent, the sharing of data has not yet reached the performance requirements of the SPS in terms of data consistency and validity. Therefore, it is necessary to build a unified platform for data sharing. The requirements for data sharing in SPSs concern the following: the range of data, the source of data, and the usage of data. The range of data can be categorized as follows:

Class 1: Static data about the physical structure and characteristics of each component of the power grid.

Class 2: Dynamic data on the operation states of the power grid. Generally, the sharing of such data should occur in real-time. The sampling interval may be in milliseconds (e.g., for protection) or seconds (e.g., for SCADA). For some noncritical measurement information, a longer sampling interval is possible (e.g., for user metering data).

Class 3: Control strategy and operation parameter settings.
Class 4: Forecast and scheduling data without high requirements for promptness.
Class 5: Historical data, including data under normal operation and recorded fault information (with an interval of nanoseconds).

The subjects of data sharing include control centers, power plants, substations, and DGs. The source and usage of data are summarized in Table 4.2. In fact, there are needs for data sharing between a control center (or substation) and users, but the effect on the SPS is negligible, so users are not listed in the table.

Table 4.2: The source and usage of data in the smart power system (CU table).

Type \ Subjects	Control centers	Power plants	Substations	Distributed generations
1	U	CU	CU	CU
2	U	C	C	C
3	C	U	U	U
4	C	U	U	U
5	CU	C	C	C

C, create; U, use; CU, both.

4.3.2 The structure and technologies of the data-sharing platform

For the above communication demands, a data-sharing platform can be built, as shown in Figure 4.2. On this platform, service applications can query, locate, and access data through a universal data access interface. After receiving a request, the data exchange platform first locates the data and calls the appropriate communication protocol. Then, the request is passed to the application that owns the data through the basic communication platform. Finally, the data, which are represented in the Common Information Model (CIM), are sent to the requesting application through the data exchange platform.

During data exchange, the independence between the data access method and the application is achieved through the universal data access interface. The independence between the data representation and the application is achieved by the CIM. The independence between data processing and data locating is achieved by the data exchange platform. Data processing and data locating share a common data-sharing mechanism, thereby improving the consistency and validity of the data.

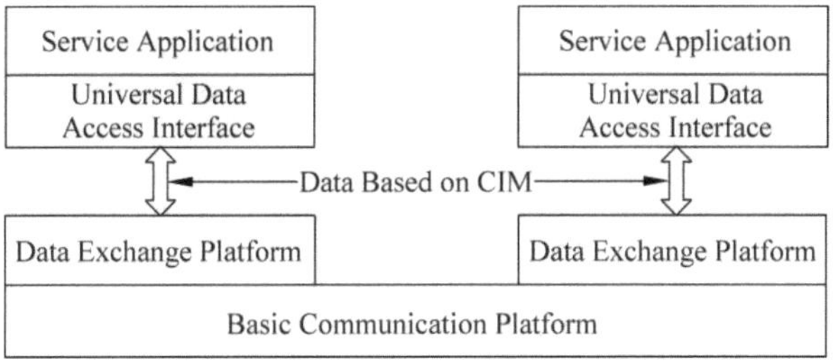

Figure 4.2: The structure of the data-sharing platform.

4.3.2.1 Common Information Model

The CIM regulates the representation of data to ensure the uniqueness of the description. In other words, the model adopts a consistent manner to describe the same object. Similar to those of the universal data access interface, its standardization efforts are reflected in two standards, namely, the CIM based on IEC 61970/61968 and the object model based on IEC 61850. The CIM is mainly intended for EMS, DMS, and other advanced applications, while IEC 61850 is targeted at process-level applications. In addition to these three standards, there are some emerging compatible expansions of standards. The information model covers almost all the application fields in the power system and can be used to manage a regional grid and the entire grid. A brief introduction is given as follows:

(1) Data model associated with an EMS/DMS application

The CIM based on IEC 61970/61968 provides a description of the typical data in the power system. A specific application generally will use only some of these classes. The information models used by EMSs are organized in packages, including the core package, domain package, load model package, measurement package, outage package, topology package, wire package, spare package, generation package, SCADA package, protection package, and energy dispatch package. In addition, a DMS application includes an asset package, user package, core-2 package, document package, ERP support package, and work package. Moreover, there is a complementary market operation package, which is mainly targeted at the US electricity market and may not necessarily be suitable for China.

(2) Process-oriented data model

With the core of IEC 61850-7, the information of process-level devices is described by using an object-oriented approach, and the main physical process-level objects (including a number of primary equipment and secondary equipment for measurement, control, and protection) and the communication network are abstracted as corresponding logical devices. The primary aim of IEC 61850 is to model process-level devices. By abstracting the substation, IEC 61850 defines approximately 80 logical nodes, which serve data objects (i.e., measurements and states in the traditional sense) and expand to IEDs. The original IEC 61850 does not consider modeling power quality. To compensate for this drawback, IEC 61850-7-3(-4) complementarily defines the relevant logical nodes, data objects, and data types for the power quality model. This work is based primarily on the assessment of power quality by using the root mean square of variance. This approach also considers other power quality indicators, such as voltage flicker, voltage dips, and frequency variation.

(3) Model related to wind generation

Similar to IEC 61850, IEC 61400 models the process information associated with wind generation [35]. The standards are meant primarily for communication between a wind turbine and other devices within the plant. IEC 61400 also considers communication with equipment and systems outside the wind plant. IEC 61400 focuses on the specific data objects associated with wind power, which are more complicated than those defined in IEC 61850.

(4) Model of distributed generation

This model mainly considers equipment for distributed power generation, including diesel generators, wind power, and solar power. The object model of DG is developed based on IEC Technical Committee 57 Working Group 17 (IEC TC57 WG17), which is based on IEC 61850 and extends the model on the characteristics of distributed energy. IEC TC57 WG17 is currently a draft standard [36].

(5) Hydro model

The hydro model aims to describe the information of the flow characteristics, the dam gates, and so on. This model is compatible with IEC 61850 and is reflected in draft standards IEC 61850-7-410.

The development of the above models is based on the concept of object modeling and the core components of IEC 61850 whenever possible. Given the demands being satisfied, these models maximize the use of existing public data classes and logical nodes. Because the standardization is carried out by different working groups, these information models have some overlap, and the coordination of these models is currently underway.

4.3.2.2 Data exchange platform

Data exchange is one of the core functions of data sharing. Therefore, a unified data exchange platform must be established. This "unity" comprises two aspects: First, the information should be uniquely identified within the system. The identification of the data used by the system is supposed to be developed and maintained by one organization. This identification is the basis for exchanging and sharing data. Second, the data should be represented uniformly. Therefore, it is necessary to establish unified, and unambiguous information representation rules in various departments and systems to standardize and normalize the classification and encoding of data.

A model of the data exchange platform is shown in Figure 4.3. For data requests from local and offsite applications, the data exchange platform has different response approaches.

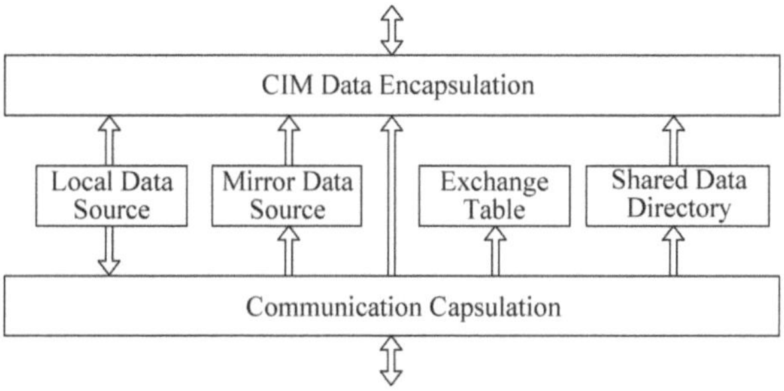

Figure 4.3: Data exchange platform.

(1) Data requests from local applications

Requested local data sources can be directly accessed. Requested offsite data need to be treated separately. If data are frequently requested, the data exchange platform can construct a local mirror of the data; this mirror is synchronized regularly. Then, accessing these data becomes no different from accessing local data. If there is no local mirror of the data, the data exchange platform should access the data source in accordance with the method written in the shared data directory, and then the data are returned for the use of the local application.

(2) Data requests from offsite applications

These requests also need to be treated separately. For data that are requested regularly for synchronization, the data exchange platform updates the data in accordance with the exchange table. For data requests that occur randomly, the data exchange platform should access local data after verifying the authority of the requests, and encapsulated data are returned following the corresponding communication protocol.

For accurate data exchanges, the synchronization of shared data must be ensured; this synchronization is the responsibility of the data exchange platform where the original data are located and should be completed based on the contents of the data exchange table. Methods of synchronization include the variable synchronous mode and timed synchronous mode. Moreover, the data exchange platform must provide some basic services, such as data directory synchronization, data exchange table synchronization, data encapsulation, and data access management.

The shared data directory provides a directory of the data that can be shared within the system and its methods of authorization and access. The update mechanism of the directory can be aggregated or distributed, but the entire data exchange system must maintain a complete data directory.

The data exchange table is used to maintain local information related to data synchronization; this information includes the external data descriptions required by local applications and the local data requirements of external systems. The synchronization of data is kept between the two sides of the exchange without the need to maintain a global data exchange table.

The encapsulation of data consists of two aspects: First, the encapsulation is based on the CIM. Second, the data exchange platform is supposed to encapsulate (or decapsulate) the data in accordance with the corresponding communication protocols when exchanging data with other platforms.

Data access management includes the update and verification of authorization. There are multiple ways to implement this type of management. For instance, authorization updates can be performed along with the synchronization of the data directory, or they can be performed asynchronously. Verification of data access can be conducted when the data are accessed or can be carried out during an inquiry of the data directory. (Consequently, unauthorized applications cannot know the relevant data directory entries.)

The data exchange platform described above is merely a conceptual model. The functions involved are only the basic ones; practical implementation would require further refinement.

4.3.2.3 Universal data access interface

A universal data access interface provides a common data access mechanism that should have the following three characteristics: independence of the specific application; independence of the form and content of the data; and support for common access patterns, such as a request/response mode and publish/subscribe mode. The universality of the access mechanism is guaranteed through the development of standards such as IEC 61970/IEC 61968/IEC 61850 [19, 36–41]. A related discussion is provided below.

For EMS applications, IEC 61970 extends and defines the following four types of data access interfaces based on the Data Access Framework (DAF), Data Access and Integration Services (DAIS), and Historical Data Access from Industrial Systems (HDAIS) data access interfaces defined by the Object Management Group (OMG).

(1) General data access (GDA)

GDA is a question and answer (Q&A)-oriented interface that supports the positioning and inquiries of structured data (including data structures and actual data). GDA provides typical database functions for reading and writing data with neutrality toward platforms and data patterns. In addition, GDA provides simple event services that inform the end clients when the data change on the server-side in a timely manner.

(2) General event subscription (GES)

GES is an interface oriented toward the publish/subscribe mode and supports hierarchical browsing of structure and instance data. GES is mainly used for application integration, in which different applications can publish and subscribe to CIM-based data through the interface provided by GES. Because GES is oriented toward the publish/subscribe mode, it naturally has nothing to do with the characteristics of developers.

(3) High-speed data access (HSDA)

HSDA is an interface that is oriented toward both the Q&A mode and the publish/subscribe mode. HSDA supports hierarchical high-speed browsing and querying of structure data and instance data. This interface focuses on the demand for high-throughput data exchange and can support the exchange of tens of thousands of data within 1 s.

(4) Time series data access (TSDA)

TSDA is also an interface that is simultaneously oriented toward both the Q&A mode and the publish/subscribe mode. This interface supports hierarchical high-speed browsing and querying of structure and instance time series data.

For applications in the DMS, although the IEC adopts the CIM, the IEC does not use the universal interface in the strict sense. Instead, the IEC defines interfaces for different application systems that are oriented toward the exchanged data. Relevant regulations on the interfaces are stated in the IEC 61968 standards and cover the applications of asset management, construction and maintenance, operation planning and optimization, network expansion planning, customer support, and meter reading and control.

In the process control, data exchange is usually a series of operations, such as sending and receiving data, reporting data triggered by an event, and keeping data logs. In response to the needs of information exchange, IEC 61850 defines a set of abstract services covering all aspects of data connection management and data collection. In addition, to flexibly accommodate various bottom-level communication protocols and modes in practical applications, IEC 61850 currently defines two mapping protocols: Generic Substation Events, which is a transfer protocol among high-speed devices, and Manufacturing Message Specification based on Transmission Control Protocol/IP.

4.3.3 Real-time data sharing with a kernel of advanced state estimation

In the operation and control of the power system, the states of the power grid are acquired in real-time; the real-time acquisition of these states is the basis for the real-time analysis and control of grid states. Due to the bulky size of modern power systems, a few measurement points are insufficient to represent the overall operation of the power system; therefore, state estimation is necessary to analyze and process the many measurements taken to obtain the whole picture of the system-wide real-time operation status.

State estimation is a relatively mature module in the EMS. Early state estimation used mostly SCADA data as data sources, thereby accomplishing the function of identifying the overall system operational status through bad data identification and cleaning. With the advancements in measurement technology, wide-area measurement systems have been promoted. The data accuracy and sampling rate have been greatly improved, yet state estimation algorithms need to be further refined. It is necessary to study and develop state estimation algorithms with high reliability, high fake detection performance, high precision, and high response speed (i.e., an advanced state estimation unit). Then, the data sharing of the genuine real-time operational states of the power system may be realized. This approach can be used for grid scheduling and dispatch.

The unit is the most demanding part in terms of promptness and accuracy in the data-sharing platform of the SPS.

4.4 Summary

The global sharing of data is one of the fundamental functions on which the optimal decision-making and coordinated control of the SPS relies. To this end, it is necessary to establish a basic communication platform and a data-sharing platform. The former is a physical transmission platform for the exchange of data, and the latter is a logical transmission platform for the integration of data.

First, this chapter describes the functions, requirements, and implementation architecture of the basic communication platform. The basic idea is to build a stable and reliable clock synchronization system and then construct a high-performance basic information transmission network. The ultimate goal is to provide data exchange and sharing hosting platform for a variety of production and application systems (i.e., a data network platform).

Second, the data-sharing platform for the SPS proposed in this chapter achieves transparency in the location and application of data while ensuring data consistency. This platform provides a solid foundation for the construction of the SPS. Additionally, a few key technical issues should be addressed in the operationalization; these issues include uniform data representation, consistent data location, and data access, real-time simulation based on mirrors, GPS-based data synchronization, and unified data access authorization management. The solutions to these problems should be started by considering the needs of the data-sharing platform to determine its overall framework. Then, research on the framework should be conducted to address specific implementation issues, such as standardization, change and interpretation mechanisms of data representation, formal methods of model transformation, and determination and interpretation of communication protocols. In addition, this chapter also notes the need to study the specific implementation of data transmission to uniformly manage data transmission links, data sending and receiving, encryption and decryption, load balancing, and so on. Research should be conducted to construct an intelligent power system and its communications infrastructure.

The last part of this chapter raises a decisive issue in the construction of the SPS; that is, traditional state estimation units are replaced with advanced state estimation units (AEUs). Hence, a conceptual reform of the design of state estimation units is required.

Chapter 5
Standard indexes for smart power system operation

5.1 Introduction

The smart power system aims to realize the “multi-indicator optimality approximating operation”. The “multi-index” mentioned here mainly refers to the three major categories of safety, quality, and economy. To realize “multi-indicator optimality approximating operation”, it is first necessary to set up operational performance indicators which define the targets and standards of power system operations. Then according to variations of such operational performance indicators, dispatchers can not only qualitatively but also quantitatively assess the state of the power system. When one or a group of operational performance indicators deviate from its standard values, and such deviations exceed allowable limitations, the current state of the power system may be identified as an unsatisfactory one. Based on the power hybrid control method, this unsatisfactory state can be further defined as an event. That is to say, an event occurs in the physical power system at this time, and proper control measures need to be applied to eliminate the event.

This chapter is cooked to introduce valuable smart power system operational performance indicators. Section 5.2 outlines the basic ideas and rules to create an index system that standardizes the power system operations. Sections 5.3 and 5.4 elaborate on the two types of important operation indicators in the smart power system:

(1) Safety indicators

Ensuring the safety of the power system is the priority of system operators. For assessing the safety of the power system comprehensively, in section 5.3, a new minimum radius method will be proposed for estimating voltage stability domains, small disturbance stability domains, and transient stability domains. By solving the minimum radius, the safety operation margin can be obtained in regards to voltage stability, small disturbance stability, and transient stability. This method may include various operating constraints easily and be implemented for online applications while having no numerical convergence issues.

(2) Performance indicators of coordinated control of interconnected power systems

By interconnecting area power grids, modern power systems gain remarkable enhancement of efficiency and reliability for the sake of increasing complexity in the system structure and operation. As the area power grids are managed by different utilities, the coordinated control among them becomes very challenging. It is necessary to derive suitable performance indicators, which can be used to not only define the responsibilities of local utilities but also assess the effectiveness of the coordinated controls. For

https://doi.org/10.1515/9783110448825-005

the coordinated active power as well as frequency controls, the performance indicator with the name of the area control error has been widely adopted in automatic generation control systems. However, for the coordinated reactive power and voltage control, few performance indicators are serving as similar standards. Filling this gap, a novel indicator is coined for multi-area voltage control applications, which will be introduced carefully in section 5.4.

5.2 Standard system of operational performance indicators

5.2.1 A standard index system

At present, there have been many works on designing operational performance indicators of power systems, which cover multiple aspects such as voltage safety, transient stability, voltage quality, operating economy, and equipment adequacy. However, these indicators are often only used to assess the power system performance from a certain angle, so they lack standardization and systematicness. Thus, it is necessary to set up a standard system of performance indicators that categorizes and layers different indicators for better understanding and controlling complicated power systems.

Establishing the standard system of performance indicators is a typical system engineering job. First, the design requirement of this index system should be clarified carefully. Then, its hierarchical structure can be outlined properly. Finally, the operation data of the power system needs to be fully utilized to update performance indicators and enhance their availability and practicality.

The purpose of establishing the standard index system is to help dispatchers and management personnel to understand the current state of the power system in a timely and comprehensive manner and identify vulnerabilities. Essentially, the index system needs to answer the following three questions, such as: "How safe is the power system? How high is the quality of its services? How efficient and valuable is the power system?" The above questions involve three objectives of the smart power system, such as long-lasting safety, high-quality service, and economical operations.

From the perspective of the hierarchical structure, the entire index system can be divided into the following three levels:

(1) Basic indicator

Basic indicators are directly collected from the power system operating data or obtained through simple calculations, which reflect the actual state of the power system directly.

(2) Advanced indicator

Advanced indicators are forged from basic indicators through data mining, statistical analysis, and other means, which reflect the mutual influence of different performance indicators.

(3) Comprehensive indicator

Comprehensive indicators are built based on basic and advanced indicators, which reflect the system-level performance of the whole power system.

5.2.2 Components of the standard index system

The standard system of performance indicators of a smart power system should be composed of three parts: the basic indicator, the advanced indicator, and the comprehensive indicator, which will be further explained below.

1. Basic indicator

(1) Basic indicators of power system state are directly related to power system safety, operation quality, and economy. These standard indicators can be further divided into real-time ones that reflect the current state of the power system and statistical ones that reflect the operating status of the power system in a certain period. Table 5.1 lists the common basic indicators that reflect the real-time and statistical state of the power system.

Table 5.1: Basic indicators of power system state.

Category 1: Safety-related indicators	Frequency safety indicator
	Voltage safety indicator
	Active power safety indicator
	Low-frequency oscillation indicator
	Transient stability indicator
Category 2: Quality-related indicators	Tie line power controllability
	Annual load loss
	Frequency quality
	Voltage quality
	Harmonic level
Category 3: Economic-related indicators	Transmission grid loss
	Hydro-energy Utilization
	Thermal power plant coal consumption
	Power generation cost
	The proportion of Wind energy, Solar energy, and small hydropower energy
	Energy-saving and emission reduction

(2) Basic indicators on the grid topology and power system equipment conditions

By analyzing features of the power grid under various operating conditions, system performance indicators such as power distribution balance, network vulnerability, and operation reliability can be obtained. Table 5.2 lists basic indicators of system topology and power equipment conditions.

Table 5.2: Basic indicators of system topology and equipment conditions.

Grid topology related indicators	Cluster coefficient
	System fragility
	Critical power transfer capacity
	Branch importance weight
	Shortest path length
	The capability of islanding operation
Load related indicators	Total load
	Load types
	Load distribution
	Load growth direction
	Bus-oriented load forecasting
Power generation related indicators	Balance of power source distributed
	The capacity of total power generation
	The capacity of power inputs from external sources
	The proportion of various types of power generations
	The growth direction of power generation

(3) The basic indicators of the automation system operation in the power system

Communication, automation, and automatic control systems play a particularly important role in smart power systems. Generally, such an automation system should be highly available, reliable, and efficient to make the smart power system fully monitored, closed-loop controlled, and well protected. Moreover, serving certain functions and tasks, each automation subsystem must also satisfy specific performance requirements and standards from which related performance indicators can be derived for example, the accuracy of the load forecast and the pass rate of the state estimation. Table 5. 3 lists some basic indicators for evaluating the automation system.

Table 5.3: Basic indicators of the automation system.

AGC	Online availability
	Ramping limits
	Command period
	Successful execution rate
AVC	Command period
	Successful execution rate
	Availability of actuators on substations
	Number Limits of daily regulations
SCADA	Sampling rate
	Accuracy of measurements
	Abundancy
	Response rapidity
State estimation	The acceptable rate of estimated states
Various Predict System	Accuracy
Operation planning	Optimality
	Robustness

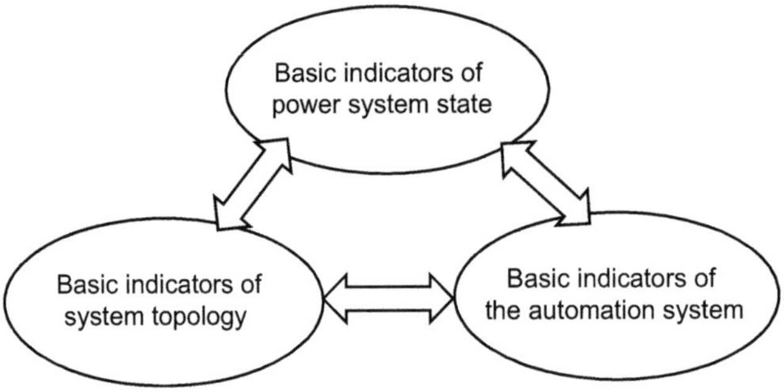

Figure 5.1: Coupling between basic indicators.

The above three types of indicators together serve as the foundation of the overall standard index system and are coupled with each other closely, as shown in Figure 5.1.

2. Advanced indicators

The basic indicators can be used for assessing the operating performance of the power system and discovering a direction to make it better. Furthermore, it is necessary to explore and analyze the mutual influence of these indicators to identify critical factors and control variables. Here, we introduce advanced indicators which are generated by data mining and statistical analyzing basic indicators. These advanced indicators help to render insights into system dynamics and form high-level operation standards.

Advanced indicators may also be correlations between different performance indicators. For example, advanced safety indicators can be developed to quantify the influences of power grid topology variations on the system stability after major disturbances. Such indicators may also be extended to consider the safety influences of other factors such as the balance of power source distribution, generation capacity, penetration rate of renewable energy sources, and load growth patterns. We will introduce some related works in section 5.3.

3. Comprehensive indicators

Facing so many basic and advanced indicators, system operators and managers may lose their focus and become confused when there are conflicts and anomalies. Under such a circumstance, comprehensive indicators become indispensable. These indicators are developed to assemble knowledge and information of basic and advanced indicators. Then, one single comprehensive indicator may give an unambiguous reflection of the status of the whole power system on a certain aspect.

4. Spatial and temporal expansion of the index system

Those aforementioned indicators can be expanded in both space and time dimensions to satisfy power system analysis requirements of different areas and different periods.

1) Expansion in the time dimension and related analysis

As a power system is a continuous dynamic system, operators and managers must understand its past operations and states to make better decisions for current regulations. Therefore, it is necessary to design performance indicators of a smart power system evaluated in different periods, which at least should include follows.

(1) Indicators of daily operations

By statistical analyses, indicators can be designed to measure the variations of the system safety, economy, and service quality each day. Meanwhile, important correlations and factors may also be defined as advanced indicators of daily operations, such as the accuracy of the daily load forecast, weather condition, and festival load patter, etc.

(2) Indicators of monthly operations

Normally, seasonal weather and social activities influence both power generations and consumption heavily. To adapt to seasonal changes, the system topology and control strategy of a power system should be planned and adjusted carefully. Therefore, monthly evaluations of system performances become very important for improving operation plans continuously. Thus, indicators of monthly operations can be developed, which also cover the system safety, economy, and service-quality features.

(3) Indicators of yearly operations

A power system always needs a long-term expansion plan, in which yearly load growths and environment policy constraints are predicted, and the optimal location and capacity of new generators and transmission lines are given. To prepare for such an expansion, an assessment of yearly operations is indispensable, which helps to discover the vulnerable and valuable parts of a power system at the same time. Thus, indicators of yearly operations are of importance. Certainly, these indicators reflect the safety, economy, and service quality of the power system. However, the economic indicators have strong weights when comprehensive evaluations of yearly operations are conducted.

2) Expansion in the space dimension and related analysis

As the interconnection between area power grids is strengthened, huge interconnected power systems have appeared in the China, USA, and Europe. However, even closed to each other, different areas may have strong variances in weather and conditions as well as social and economic features. Consequently, every power system would be unique regarding its topology, load pattern, power source composition, and operation strategy. Then, it becomes reasonable to create specific basic indicators for area power systems by modifying the standard ones used by the whole interconnected power system.

Moreover, the hierarchical dispatch mechanism is widely adopted by the interconnected power system. Area dispatch centers carry out system performance assessments periodically. To guarantee the fairness and transparency of such assessments, advanced and comprehensive indicators of area power system operations should be standard and comparable regarding their physical meanings and evaluation algorithms.

5.2.3 Workflow of indicator computations

Figure 5.2 illustrates a general workflow of indicator computations. Of course, the computation procedure for updating a certain indicator varies remarkably. This general workflow defines necessary steps to extract critical information from power system states as follows.

(1) Data acquisition

The first step is to obtain the fundamental data of power system operations, including ① the topology of the power grid, parameters and states of power equipment from SCADA, PMU, and EMS systems; ② the price of electricity products, short and long term energy contracts, and market restrictions from electricity trading systems, ③ operation states of various automation and information components.

(2) Data preparation

Within the huge amount of power system data, errors and mismatches inevitably result from unreliable data measuring and transferring. Directly using such raw data, system indicators may not reflect the real status of the power system leading to risky control decisions. Therefore, the preprocessing of the raw data is an important step, during which state estimations are carried out to screen out bad measurements and obtain trustable system states.

(3) Computation of basic indicators

Most basic indicators may be directly extracted from the measured and estimated system states. Computation related to basic indicators is mainly associated with cumulations and comparisons.

(4) Computation of advanced indicators

It is a critical step to compute the values of advanced indicators for building up the standard index system. Compared to basic indicators, advanced indicators are generated by advanced statistical techniques such as data mining and knowledge discoveries. Through processing basic indicators, deep insights of correlations between the network topology, equipment status, automation flows, and power system states.

(5) Computation of comprehensive indicators

Comprehensive indicators recover the optimality of power system operations, from comparing performances of different power systems or the performance of the same power system in different periods. Especially, comprehensive indicators may adopt fuzz classifications, where the power system status is identified as some macro state abstractly representing system features and trends. Thus, comprehensive indicators can aggregate information contained in basic and advanced indicators and give a high-level view of the whole power system.

(6) Visualization of indicators

Through visualizing indicators regarding their spatial and temporal characteristics, data, knowledge, and boundaries of power system operations can be interpreted into human-friendly information, which supports multi-objective approximate optimal operations. Therefore, visualizations of indicators are also very important as such techniques enable the closed-loop control of the power system with human operators.

(7) Application of indicators

The main functions of the aforementioned indicators are to identify events and guide the optimal control decisions. When one or multiple performance indicators are exceeding acceptable limits defined for them separately, an event may be identified, which represents an unsatisfied system state. After that, suitable control commands will be generated by optimizations using the corresponding performance indicators as objectives or constraints.

5.3 Safety indicators

Among all performance indicators, the security indicators are evaluated with the first priority. As is well known, many factors affect the safety of a power system; these factors include different types of disturbances, different disturbance intensities and fault locations, and different load growth directions and speeds. In addition, assessing the operational safety of power systems often involves very complex computational processes. The following introduces and describes several algorithms for calculating dynamic indicators of operational safety of power systems.

In theory, given a specific disturbance as well as characteristics of it (given set of changes in system operating parameters, such as the load and power generation variations), a numerical calculation method can be used to find the corresponding safety domain boundary of the power system. The fault set is usually determined by the N-1 or N-2 principle and can also be customized by the user. Here, three types of power system security domains are considered: the static voltage security domain, small-disturbance security domain, and transient security domain.

Static voltage safety domain and small-disturbance security domain can be formulated as solvable region in the parameter space of high-dimensional nonlinear equations. Currently, it is difficult to analytically describe the boundary of the static voltage safety domain and small-disturbance security domain. The transient safety domain problem of a power system in the injected space is mathematically equal to the parametric attraction region of high-dimensional nonlinear differential-algebraic equations (DAEs) at a certain equilibrium point. It is also very difficult to analytically solve the transient security domain as well as its boundaries.

For solving the load safety margin, the available transmission power limits and other safety indicators, power growth directions of loads and power generations are specified in the simulation. Then, along with these directions, in accordance with a certain step, the system load power and power generation are gradually increased until the voltage instability, small-disturbance instability, transient instability, or power flow solution violating operational constraints.

That is, the specified injection growth direction is used to find the security domain boundaries that constitute the system. In theory, a search for all possible nodal power growth directions can figure out the security domain boundaries in the injection space. However, this exhaustive search is computationally expensive and difficult to use online.

The natural idea is to find the shortest path from current state $X(t)$ to the safe domain boundary. Here, the security domain boundary exists for a particular group or a specific set of disturbances.

Based on the sensitivity of the safety indicator to the nodal injection power, this section proposes a heuristic algorithm for estimating the shortest distance to the security domain boundary. The principles of this method are applicable to the voltage safety domain, small-disturbance security domain, and transient security domain in the injection space. Because this method is a heuristic method, there is no numerical convergence problem, and it is easy to consider various operating constraints.

5.3.1 Minimum radius of the voltage safety domain and its calculation method

5.3.1.1 Principle

The boundary C($\boldsymbol{R}^n$) of the voltage safety domain for a certain disturbance F in the injection space is composed of all $\boldsymbol{P}^*$ points; that is,

$$\mathrm{C} = \{\boldsymbol{P}^* | \boldsymbol{f}(\boldsymbol{X}, \boldsymbol{P}) = 0, \boldsymbol{g}(\boldsymbol{X}, \boldsymbol{P}) \le 0, \boldsymbol{P} \in \boldsymbol{R}^n\} \tag{5.1}$$

where $\boldsymbol{P}^*$ is the load power vector for the node, $\boldsymbol{X}$ is the system state vector except for the injected power $\boldsymbol{P}$, $\boldsymbol{f}(\boldsymbol{X}, \boldsymbol{P}) = 0$ is the power flow equation with consideration of the disturbance F, and $\boldsymbol{g}(\boldsymbol{X}, \boldsymbol{P}) \le 0$ is the operation constraint. Under the normal operation condition, the voltage safety domain is composed of all the nodal injection vectors that satisfy the power flow equation and the operational constraints. The voltage safety field Ω can be expressed as follows:

$$\Omega = \{\boldsymbol{P} \,|||\, \boldsymbol{P} - \boldsymbol{P}^0|| \le \boldsymbol{r}^*,\ \boldsymbol{r}^* = ||\boldsymbol{P}^* - \boldsymbol{P}^0|| \tag{5.2}$$

where $\boldsymbol{P}^0$ is the power injection vector at T_0 moment; P is the power injection vector of the next time moment; it is clear that $\boldsymbol{P} - \boldsymbol{P}^0$, which is the increment vector of the power injection (see Figure 5.2); and $\boldsymbol{P}^* - \boldsymbol{P}^0 = \Delta\boldsymbol{P}$ represents the power injection vector along the most dangerous direction. The norm of the vector $\Delta\boldsymbol{P}$ is that is, the modulus of the vector is represented by the scalar r as follows:

$$||\Delta\boldsymbol{P}|| = ||\boldsymbol{P} - \boldsymbol{P}^0|| = r \tag{5.3}$$

As above, r^* represents the distance from the current operation point at T_0 to the safety boundary Ω along the direction as $\Delta\boldsymbol{P} = \boldsymbol{P}^* - \boldsymbol{P}^0$. As shown in Figure 5.3, r^* can also be called the minimum radius (distance) in the $\boldsymbol{P}^* - \boldsymbol{P}^0$ direction with point B (point $\boldsymbol{X}^0$) as the center or the maximum allowable radius.

$\Delta\boldsymbol{P} = \boldsymbol{P}(T_0 + \Delta t) - \boldsymbol{P}^0(T_0)$. The map on the plane

The problem is now clear; that is, the voltage safety conditions are determined by

$$\boldsymbol{r} = ||\boldsymbol{P} - \boldsymbol{P}^0|| \le \boldsymbol{r}^* \tag{5.4}$$

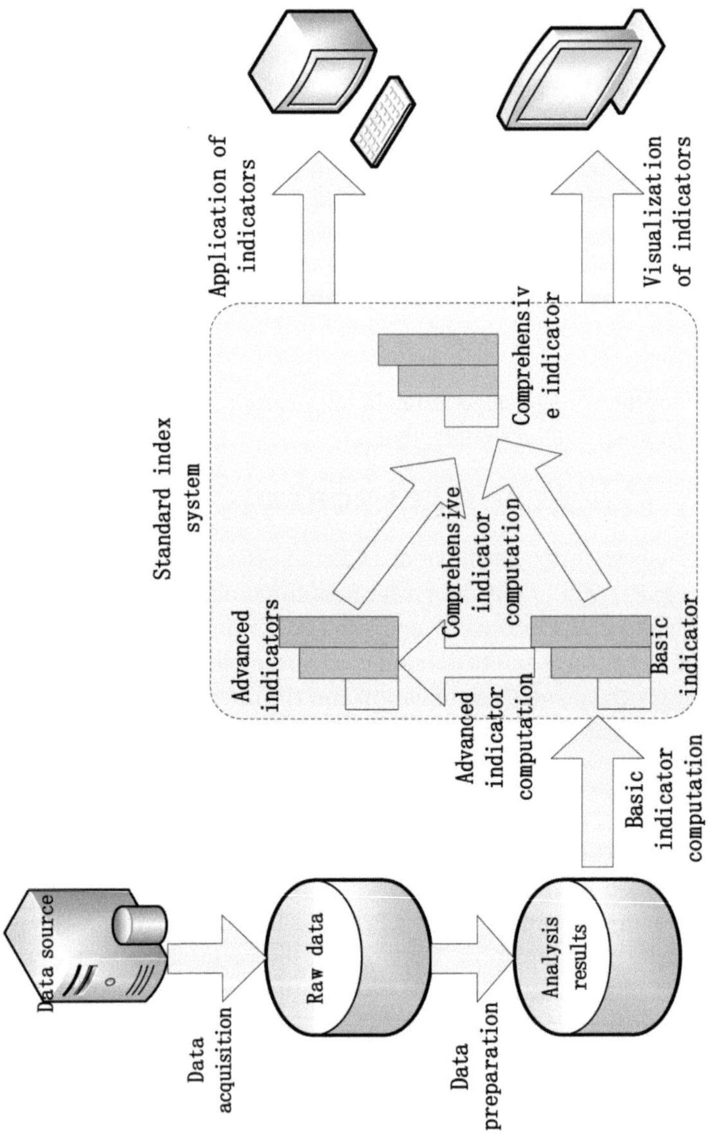

Figure 5.2: Increment of power injection vector.

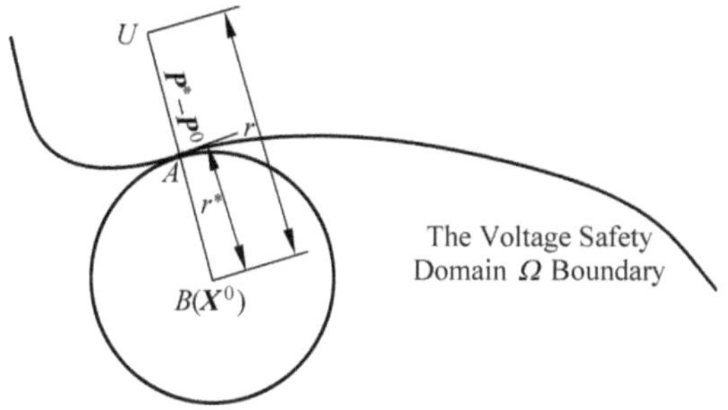

Figure 5.3: Diagram of the voltage safety domain Ω boundary and the maximum allowable radius r^* (the U-point system voltage shown in the figure is in an unsafe state).

It can be seen that criterion (5.4) is a sufficient condition for the safety of the power system and can be written in the following concise form:

$$\overline{\Omega} = \{P|||P - P^0|| \le r^*, ||P_i|| \ge ||P_i^0||\} \quad \{i = 1, 2, \ldots, n\} \tag{5.5}$$

where i is the index of injection nodes, and $\overline{\Omega}$ is the voltage safety domain. The conditions in the above equation $||\boldsymbol{P}_i|| \ge ||\boldsymbol{P}_i^0||$ are not theoretically needed but rather are for engineering practice. When voltage safety issues are concerned, increments of loads are much more dangerous than reductions of loads.

Now, we will address the problem in which a power system has many nodes; while the load on each node increases, there could be infinite directions of injection power growth in theory. For solving one or a limited number of the most dangerous directions of injection power growths, the computational effort tends to be unacceptably large for online applications. Thus, several critical nodes and a limited number of dangerous directions can be derived based on the experience and knowledge the dispatching experts of the power system. As a result, the related computation is not excessive.

The following is a heuristic method based on the voltage stability quantization index for estimating r^*.

The eigenvalue of the Jacobian matrix of the power flow equation can represent the degree of singularity. When the minimum modulus of these eigenvalues is close to zero, the power system approaches voltage instability. Therefore, the minimum modulus eigenvalue of the Jacobian matrix can be chosen as the quantization index η of the voltage stability. Let the minimum modulus eigenvalue of the Jacobian matrix of the power flow equation at the current operating point be

$$\eta^0 = \lambda^0_{\min} = \min_i \left(\lambda_i\left(\boldsymbol{J}\left(\boldsymbol{P}^0\right)\right)\right) \tag{5.6}$$

where $\boldsymbol{J}(\boldsymbol{P}_0) = \frac{\partial \boldsymbol{f}}{\partial \boldsymbol{x}}\big|_{\boldsymbol{P}=\boldsymbol{P}^0}$, $\boldsymbol{P}^0$ is the nodal injection power at time T_0, $\boldsymbol{P}^0 = \left[\boldsymbol{P}_G^0, \boldsymbol{P}_L^0\right]^T$, $\boldsymbol{P}_G^0 = \left[P_{G,1}^0, P_{G,2}^0, \ldots, P_{G,m}^0\right]^T$ is the power generation vector, and $\boldsymbol{P}_L^0 = \left[P_{L,1}^0, P_{L,2}^0, \ldots, P_{L,n}^0\right]^T$ is the load demand vector. Then, the sensitivity vector of the voltage stability index regarding nodal power injections can be presented as follows:

$$S^0 = [S_G^0, S_L^0]^T$$
$$= \left[\frac{\partial\eta^0}{\partial P_{G,1}^0}, \frac{\partial\eta^0}{\partial P_{G,2}^0}, \ldots, \frac{\partial\eta^0}{\partial P_{G,m}^0}; \frac{\partial\eta^0}{\partial P_{L,1}^0}, \frac{\partial\eta^0}{\partial P_{L,2}^0}, \ldots, \frac{\partial\eta^0}{\partial P_{L,n}^0}\right]^T \tag{5.7}$$

where $\frac{\partial\eta^0}{\partial P_{G,k}^0}$ and $\frac{\partial\eta^0}{\partial P_{L,l}^0}$ are items of vectors S_G^0 and S_L^0, respectively, which are negative and have the largest absolute values as follows:

$$\frac{\partial\eta^0}{\partial P_{G,k}^0} = \min_i \left(\frac{\partial\eta^0}{\partial P_{G,i}^0}\right) \tag{5.8}$$

$$\frac{\partial\eta^0}{\partial P_{L,l}^0} = \min_j \left(\frac{\partial\eta^0}{\partial P_{L,j}^0}\right) \tag{5.9}$$

According to equations (5.8) and (5.9), at T_0, the most dangerous growth of nodal power injections should be related to the kth generation unit and the lth load, respectively. The following two vectors $I_{G,W}^*$, and $I_{L,W}^*$ the most dangerous power injection growth direction, as follows:

$$I_{G,W}^* = [0, \ldots, 0, k, 0, \ldots, 0]^T \tag{5.10}$$

$$I_{L,W}^* = [0, \ldots, 0, l, 0, \ldots, 0]^T \tag{5.11}$$

$$I_n^0 = \left[I_{G\cdot W}^0, I_{L\cdot W}^0\right]^T \tag{5.12}$$

A step length is set as h, which is used to increase the power generation and load demands proportionally. After a step-growth, the nodal power injections of the whole power system become

$$P^1 = P^0 + k^0 \cdot h \tag{5.13}$$

where h is small enough, $P^1 = P(T_0 + \Delta t)$, and k^0 represent the power increment along the most dangerous growth direction.

Then, with $P = P^1$ used as a starting point, the sensitivity vector of the voltage stability quantization index on node S^1 is calculated again, and then the most dangerous growth direction S^1 at P^1 is solved based on k^1. The above process is repeated until the power system loses voltage stability. When the power injection increases by step $u+1$, the system is unstable, and the critical power generation and load level of the system is

$$P^u = P^0 + h\sum_{i=0}^{u} k^i \tag{5.14}$$

Along the most dangerous growth direction, the power generation and load margins are

$$\boldsymbol{P}_{\text{gen, margin}} = ||\boldsymbol{P}_G^u|| - ||\boldsymbol{P}_G^0|| \tag{5.15}$$

$$\boldsymbol{P}_{\text{load, margin}} = ||\boldsymbol{P}_L^u|| - ||\boldsymbol{P}_L^0|| \tag{5.16}$$

The minimum radius of the voltage safety domain in the injection space (see Figure 5.3) is

$$r^* = ||\boldsymbol{P}^u - \boldsymbol{P}^0|| \tag{5.17}$$

where $|| \bullet ||$ denotes the norm, $\boldsymbol{P}_G^{cr} = \boldsymbol{P}_{\text{gen, margin}} + ||\boldsymbol{P}_G^0||$ is the critical power generation level, and $\boldsymbol{P}_L^{\text{cr}} = \boldsymbol{P}_{\text{load, margin}} + ||\boldsymbol{P}_L^0||$ is the critical load level. The stability margins shown in equations (5.15) and (5.16) are the minimum stability margins for the various possible power generation and load growth direction; as long as the total load growth of the system does not exceed the limit given by (5.16), the system is voltage stable. The power generation and load stability margins shown in equations (5.15) and (5.16) can be used as important indicators of safety monitoring. The above method can approximate the nearest boundary of the voltage safety domain, during which there are no numerical convergence issues.

5.3.1.2 The sensitivity of the Jacobian matrix feature root to node injection

The power flow equation can be

$$\boldsymbol{f}(\boldsymbol{X}) = \boldsymbol{P} \tag{5.18}$$

where $\boldsymbol{X}$ is the node voltage amplitude, and phase angle and P is the node power injection vector. The minimum modulus eigenvalue of the Jacobian matrix $\frac{\partial f}{\partial \boldsymbol{X}}$ at the solution $\boldsymbol{X} = \boldsymbol{X}^*$ is λ, the corresponding left eigenvector is w, and the right eigenvector is v. Then,

$$\frac{\partial \lambda}{\partial \boldsymbol{P}} = \frac{\partial \lambda}{\partial \boldsymbol{X}} \cdot \frac{\partial \boldsymbol{X}}{\partial \boldsymbol{P}} = \frac{\partial \lambda}{\partial \boldsymbol{X}} \left[\frac{\partial \boldsymbol{f}}{\partial \boldsymbol{X}}\right]^{-1} \tag{5.19}$$

where $\frac{\partial \lambda}{\partial \boldsymbol{X}} = \left[\frac{\partial \lambda}{\partial x_1}, \frac{\partial \lambda}{\partial x_2}, \ldots, \frac{\partial \lambda}{\partial x_n}\right]$. The function of the element x in the Jacobian matrix can be derived from the sensitivity of the eigenvalues to the matrix parameter [42], as below,

$$\frac{\partial \lambda}{\partial x_i} = \frac{\boldsymbol{w}^T \frac{\partial^2 f}{\partial \boldsymbol{X} \partial x_i} \boldsymbol{v}}{\boldsymbol{w}^T \boldsymbol{v}} \tag{5.20}$$

Equations (5.19) and (5.20) can be used to find the sensitivity of the Jacobian matrix eigenvalues at $\boldsymbol{X} = \boldsymbol{X}^*$ to the injected power of the nodes.

5.3.2 Minimum radius of the small-disturbance safety domain and its calculation method

5.3.2.1 Principle

The small-disturbance domain in the injection space is defined as

$$\Omega = \{\boldsymbol{P}|\xi_{\min}(\boldsymbol{A}_{sys}(\boldsymbol{P})) \geq \varepsilon, f(\boldsymbol{X},\boldsymbol{P}) = 0, g(\boldsymbol{X},\boldsymbol{P}) \leq 0, \boldsymbol{P} \in \boldsymbol{R}^n\} \tag{5.21}$$

where P is the power injection vector of the node in any direction, $\boldsymbol{A}_{\text{sys}}(\boldsymbol{P})$ is the system matrix of the dynamic model of power system linearization (see equation (5.29)), $\xi(\boldsymbol{A}_{\text{sys}}(\boldsymbol{P}))_{\min}$ is the damping ratio of the dominant mode, $\varepsilon > 0$ is the threshold for setting the minimum damping ratio, and $\boldsymbol{g}(\boldsymbol{x},\boldsymbol{P}) \leq 0$ is the constraint inequality of system operations. The small-disturbance safety domain Ω consists of all possible nodal power injections that make the dominant system mode greater than the damping ratio, and the solution of the power flow equation satisfies the operation constraint. Then, Ω can be expressed as below,

$$\Omega = \{\boldsymbol{P}|||\boldsymbol{P} - \boldsymbol{P}^0|| \leq \boldsymbol{r}^*\} \tag{5.22}$$

The symbols used in the formulas are the same as those used in Section 5.3.1 and are not repeated. The eigenvalue equation and the operation constraint are a set of linear and nonlinear equations. Similar to that described in Section 5.3.1, there exits the most dangerous direction of the nodal power injection growths. In this direction, from the point $\boldsymbol{P}^0$ in the injection space, the distance to the boundary of Ω is the shortest, as shown in Figure 5.3; that is, $\boldsymbol{P} = \boldsymbol{P}^*$ makes

$$r^* = r^*(\boldsymbol{P}^*) = \min_{\boldsymbol{P}} r(\boldsymbol{P}) \tag{5.23}$$

Finding the maximum allowable radius r^* of the small-disturbance security domain is significant for monitoring and forecasting small disturbance stability issues, especially for low-frequency oscillations. Usually, in the study of low-frequency oscillation problems, the main concerns are the margins of power generation and load in the most dangerous direction, so the safety domain can be further outlined as

$$\overline{\Omega} = \{\boldsymbol{P}|||\boldsymbol{P} - \boldsymbol{P}^0|| \leq \boldsymbol{r}^*, ||\boldsymbol{P}_i|| \geq ||\boldsymbol{P}_i^0||\} \quad \{i = 1, 2, \ldots, n\} \tag{5.24}$$

where i is the node index.

Similar to the approach described in Section 5.3.1, to reduce the computational effort, we can develop a heuristic algorithm disturbance for estimating the sensitivity of nodal power injection regarding the small-disturbance safety index.

The damping ratio of the dominant mode of a power system indicates the distance from the current system state to the small-disturbance safety domain boundary. Generally, if the damping ratio of the dominant mode is close to zero, the system tends to be unstable even facing small disturbances. Thus, the damping ratio of the dominant mode is chosen as the quantization index η of the small-disturbance safety, which can be evaluated as below,

$$\eta^0 = \xi^0_{\min} = \min_i \left(\xi_i\left(\boldsymbol{A}_{\text{sys}}\left(\boldsymbol{P}^0\right)\right)\right) \tag{5.25}$$

where $\boldsymbol{P}^0_G = \left[P^0_{G,1}, P^0_{G,2}, \ldots, P^0_{G,m}\right]^T$ is the nodal power injection vector $\boldsymbol{P}^0$ at $\boldsymbol{T}_0$, $\boldsymbol{A}_{\text{sys}}\left(\boldsymbol{P}^0\right)$, is the coefficient matrix of the linear dynamic model of the power, system $\boldsymbol{P}^0_G = \left[P^0_{G,1}, P^0_{G,2}, \ldots, P^0_{G,m}\right]^T$ is power generation vector, and $\boldsymbol{P}^0_L = \left[P^0_{L,1}, P^0_{L,2}, \ldots, P^0_{L,n}\right]^T$ is the load demand vector.

Then, the sensitivity vector of the stability quantization index to the nodal power injection can be expressed as

$$\boldsymbol{S}^0 = \left[\boldsymbol{S}^0_G, \boldsymbol{S}^0_L\right]^T = \left[\frac{\partial \eta^0}{\partial P^0_{G,1}}, \frac{\partial \eta^0}{\partial P^0_{G,2}}, \ldots, \frac{\partial \eta^0}{\partial P^0_{G,m}}; \frac{\partial \eta^0}{\partial P^0_{L,1}}, \frac{\partial \eta^0}{\partial P^0_{L,2}}, \ldots, \frac{\partial \eta^0}{\partial P^0_{L,n}}\right]^T \tag{5.26}$$

An iterative search can be carried out using the above sensitivity vector to determine the minimum load margin considering the small-disturbance stability constraint. The search algorithm is similar to the one introduced before for seeking the minimum load margin constrained by the voltage stability. To save pages, it is not elaborated here.

5.3.2.2 The sensitivity of the dominant mode damping ratio to node injection

A mathematical model of a power system can be written in the following DAE form:

$$\begin{cases} \dot{\mathbf{x}} = f(X, Y, P) \\ \mathbf{0} = g(\boldsymbol{X}, \boldsymbol{Y}, \boldsymbol{P}) \end{cases} \tag{5.27}$$

where $\boldsymbol{x}$ is the state variable, $\boldsymbol{y}$ is the output variable, and P is the nodal power injection.

To make the above equation linearized, we have

$$\begin{cases} \Delta\dot{\mathbf{x}} = \boldsymbol{f}_x(\boldsymbol{P})\Delta X + \boldsymbol{f}_y(\boldsymbol{P})\Delta Y \\ \boldsymbol{g}_x(P)\Delta\mathbf{x} + \boldsymbol{g}_y(P)\Delta\mathbf{y} = \mathbf{0} \end{cases} \tag{5.28}$$

where $\boldsymbol{f}_x = \frac{\partial f}{\partial X}$ and, depending on the nodal injection power $\boldsymbol{P}$, $\boldsymbol{f}_x(\boldsymbol{P})$ denotes the Jacobian matrix of $\boldsymbol{f}$ for $\boldsymbol{x}$.

The output variable y is eliminated using the second expression in equation (5.28) as follows:

$$\Delta\dot{\mathbf{x}} = \left[\boldsymbol{f}_X(\boldsymbol{P}) - \boldsymbol{f}_Y(\boldsymbol{P})(\boldsymbol{g}_Y(\boldsymbol{P}))^{-1}\boldsymbol{g}_X(\boldsymbol{P})\right]\Delta\boldsymbol{X} = \mathbf{A}_{\text{sys}}(\boldsymbol{P})\Delta\boldsymbol{X} \tag{5.29}$$

where, depending on the nodal injection $\boldsymbol{P}$, $\boldsymbol{A}_{\text{sys}}(\boldsymbol{P})$ is the system matrix of the power system linearization model. Since the relationship between $\boldsymbol{A}_{\text{sys}}(\boldsymbol{P})$ and the nodal injection is too complex, it is difficult to analyze the sensitivity of the dominated mode damping ratio to $\boldsymbol{P}$, which can be obtained by the following numerical perturbation method:

$$\frac{\partial \xi_{\min}}{\partial P_i} = \frac{\xi_{\min}(\mathbf{P} + \mathrm{I}_i h_s) - \xi_{\min}(\mathbf{P})}{h_s}, \quad i = 1, 2, \ldots, n \tag{5.30}$$

where $\xi_{\min}$ is the damping ratio of the dominant mode $\boldsymbol{A}_{\mathrm{sys}}(\boldsymbol{P})$; h_s is the perturbation step, and I_i is the unit vector, where the subscript i indicates that its ith component is 1 to determine its direction of growth.

5.3.3 Minimum radius of the transient safety domain and its numerical approximation

5.3.3.1 Principle

For a particular fault F, the transient security domain in the injection space is defined as

$$\Omega = \{\boldsymbol{P} | \eta(\boldsymbol{P}) \geq \varepsilon, \boldsymbol{g}(\boldsymbol{X}, \boldsymbol{P}) \leq 0, \boldsymbol{P} \in \boldsymbol{R}^n\} \tag{5.31}$$

where $\boldsymbol{P}$ is the node injection vector, $\eta(\boldsymbol{P})$ is the transient stability index under fault F, and $\boldsymbol{g}(\boldsymbol{x}, \boldsymbol{P}) \leq 0$ is the operation constraint (including before and after failure).

The transient security domain Ω is composed of all possible nodal power injections determining the system states which are able to maintain transient stability after fault F and can be expressed as follows:

$$\Omega = \{\boldsymbol{P} | \|\boldsymbol{P} - \boldsymbol{P}^0\| \leq \boldsymbol{r}^*(\boldsymbol{P}^* - \boldsymbol{P}^0) = \boldsymbol{r}^*(\boldsymbol{P}^*)\} \tag{5.32}$$

The meanings of the symbols in the formula are similar to those used in Section 5.3.2 and are not repeated. The scalar function $\boldsymbol{r}^*(\boldsymbol{P}^* - \boldsymbol{P}^0) = \boldsymbol{r}^*(\boldsymbol{P}^*)$ represents the radius of a circle centered at $\boldsymbol{P}^0$ along the direction as $\boldsymbol{P}^* - \boldsymbol{P}^0$ in the domain Ω is.

Both theory and practice tell us that there is a dangerous direction of nodal power injection growths that weaken the transient security of a power system terribly. In this direction, from $\boldsymbol{P}^0$ to the boundary of Ω, the shortest distance can be estimated as below.

$$r^* = r(\boldsymbol{P}^*) = \min_P r(\boldsymbol{P}) \tag{5.33}$$

where r^* represents the minimum permissible transient stability margin in the nodal injection space. Therefore, for a given power system with the initial nodal power injection as P^0, it can endure at least an increment of nodal power injection as r^* without losing stability. Such an increment can happen in the most dangerous direction while allocating the total r^* injection power on all nodes arbitrarily. Generally, we care only about the transient stability margin in the direction defined by the predicted load growth and corresponding power generation regulation, so the domain Ω may be further limited to

$$\bar{\Omega} = \{P|||P - P^0|| \leq r^*, ||P_i|| \geq ||P_i^0||,\} \qquad \{i = 1, 2, \ldots, n\} \tag{5.34}$$

where i is the number of nodes with power injections.

Currently, many analytical methods (such as BCU, EEAC, PEBS, and hybrid methods (HMs)) can provide transient stability quantitative indicators. Here, an HM is used to estimate the power system transient stability quantitatively.

In this HM, the time-domain simulation is carried out to give system post-fault trajectories, upon which the transient stability quantization indicator is evaluated. It combines the advantages of time-domain simulation and analytical methods (energy function methods). The time-domain simulation may adopt various and complicated system models and describe system dynamics accurately, while an analytical method can provide a system stability indicator. Here, a new hybrid method is developed, which takes the synthetic energy of synchronous machines as an indicator of system transient stability. Thus, in the following context, this method is abbreviated as SHM.

In a synchronous coordinate system, the rotor angle, rotor speed, and rotation inertia of generator i are denoted as δ_i, ω_i, and M_i, respectively, and $M_T = \sum_{i=1}^{n} M_i$; then, the rotor angle and speed of the inertia center are

$$\delta_{coi} = \frac{1}{M_T} \sum_{i=1}^{n} M_i \delta_i$$

$$\omega_{coi} = \frac{1}{M_T} \sum_{i=1}^{n} M_i \omega_i$$

In an inertial center coordinate system, the kinetic energy $V_{KE,i(t)}$ and the potential energy $V_{PE,i(t)}$ of generator i at moment t in the transient dynamic can be calculated by

$$V_{KE,i(t)} = \frac{1}{2} M_i \tilde{\omega}_i^2(t) \tag{5.35}$$

$$V_{PE,i}(t) = \int_{\tilde{\delta}_{i,s}}^{\tilde{\delta}_i} \tilde{\omega}_i d\tilde{\delta}_i \tag{5.36}$$

where $\tilde{\delta}_i = \delta_i - \delta_{wi}$ and $\tilde{\omega}_i = \omega_i - w_{wi}$ are the rotor angle and speed, respectively, of generator i in the inertial center coordinate system; and $(\tilde{\delta})_{i,s}$ is the stable equilibrium point after a fault in the inertial center coordinate system.

After the fault is cleared, the excess kinetic energy of the power generation unit will gradually be converted into potential energy. Due to the total energy conservation after the fault, the potential energy reaches the maximum value when the kinetic energy reaches the minimum value. For any power generation unit, it loses stability (synchronization) if the excess kinetic energy cannot be completely converted into potential energy. A critical power generation unit refers to the first one losing synchronization after the fault.

For the unstable scenario, the SHM gives a quantitative indicator of transient stability as follows:

$$\eta = -V_{\mathrm{KE},i}^{\min}$$

where $V_{\mathrm{KE},i}^{\min}$ is the minimum kinetic energy of critical generator i after a major fault. The reason for choosing a negative value of $V_{\mathrm{KE},i}^{\min}$ as the quantitative indicator of stability is that after the fault, the excess kinetic energy of the generator cannot be transformed into potential energy; that is, the kinetic energy $V_{\mathrm{KE},i}^{\min}>0$ when the generator is in its "potential bowl" for the critical steady-state.

For the stable scenario, the SHM also provide a quantitative indicator of transient stability as follows:

$$\eta = V_{\mathrm{PE},i}^{\max} - V_{\mathrm{PE},i}(t_{cl}) - V_{\mathrm{KE},i}(t_{cl}) \tag{5.37}$$

where t_{cl} is the clear fault time; $V_{\mathrm{PE},i}(t_{cl})$ and $V_{\mathrm{KE},i}(t_{cl})$ are the potential energy and kinetic energy of generator i, respectively; and $V_{\mathrm{PE},i}^{\max}$ is the maximum potential energy of the generator i during a constant fault, which can be evaluated as follows.

Do not clear the fault utile $t = t_{cl,\,critical.}$ Set an appropriate $t_{cl,\,critical}$ so that the kinetic energy $V_{\mathrm{KE},i}^{\min} = 0$ (i.e., the critical steady-state) when the generator i moves to its "potential bowl" edge with the maximum potential energy $V^{max}{}_{PE,\,i.}$

Additionally, after setting an operation condition with the nodal power injection $\boldsymbol{P}^0$, the corresponding SHM stability margin under a fault F is

$$\eta^0 = \eta\left(\boldsymbol{P}^0\right) \tag{5.38}$$

Then, the sensitivity of the power injections to the transient stability quantitative indicator can be presented as follows:

$$\boldsymbol{S}^0 = \left[\boldsymbol{S}_G^0, \boldsymbol{S}_L^0\right]^T = \left[\frac{\partial\eta^0}{\partial P_{G,1}^0}, \frac{\partial\eta^0}{\partial P_{G,2}^0}, \ldots, \frac{\partial\eta^0}{\partial P_{G,m}^0}; \frac{\partial\eta^0}{\partial P_{L,1}^0}, \frac{\partial\eta^0}{\partial P_{L,2}^0}, \ldots, \frac{\partial\eta^0}{\partial P_{L,n}^0}\right]^T \tag{5.39}$$

An iterative algorithm can be used to seek the minimum load margin constrained by the transient stability using the above sensitivity formulation. This solution is similar to the one used for searching the minimum load margin considering the voltage stability constraints.

5.3.3.2 The sensitivity of the energy margin to nodal power injection

It is very difficult to solve the sensitivity defined in (5.39) analytically. Here, a numerical perturbation method is used to estimate this sensitivity approximately.

$$\frac{\partial \eta}{\partial P_i} = \frac{\eta(\boldsymbol{P}+\boldsymbol{I}_i h_i)-\eta(\boldsymbol{P})}{h_s}, \qquad i=1,2,\ldots,n \tag{5.40}$$

where η is the quantitative indicator of the system transient stability provided by the SHM; $\boldsymbol{P}$ is the nodal power injection vector; h_s is the perturbation step; and $\boldsymbol{I}_i$ is the unit vector, where its i component is 1 to determine its growth direction.

5.3.4 Numerical studies

5.3.4.1 An example of the shortest radius of the voltage safety domain

The IEEE 30-node system, as shown in Figure 5.4, is used in the following tests. The initial power generation is set as 165.67 MW, while the initial load is 164.3 MW. The minimum modulus of the Jacobian matrix eigenvalues is 0.2179. According to the sensitivity of the minimum modulus eigenvalue of the Jacobian matrix to node active power injection, the load and power generation growth that is most unfavorable to the system voltage stability in the initial state are the load growth of node 21 and the power generation growth of node 25.

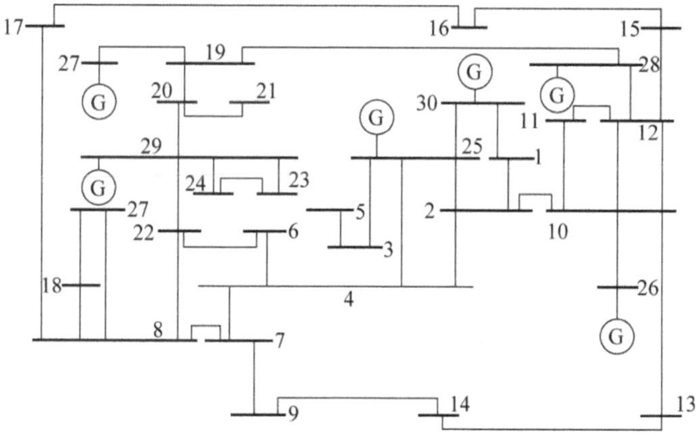

Figure 5.4: IEEE 30-bus system.

The maximum load margin is solved as 31 MW, which is approximately 18% of the initial load, calculated by 32 iterations using the method introduced in this book. This result is of importance because that no matter what proportion of the increased 31 MW load is distributed to load nodes and which generator supplies the increased 31 MW load, the system remains to be voltage stable certainly. The obtained minimum load margin can be used as an important indicator for monitoring the voltage stability level.

For investigating the influence of the operating constraints, the shortest radius of the voltage safety domain is solved under different voltage constraints as shown in

Table 5.4, where the power generation and load increase step size is set as 1 MW, and the minimum voltage constraints of all nodes are gradually tightened from 0.90 to 0.95 (the voltage of the PV node is all taken as 1.0). Obviously, with the tightening of voltage constraints, the shortest radius of the voltage safety domain should also shrink.

Table 5.4: The minimum radius of the voltage safety domain.

Voltage constraint (pu)	Minimum radius/MW	Voltage constraint (pu)	Minimum radius/MW
0.90	10	0.93	6
0.91	9	0.94	5
0.92	7	0.95	4

5.3.4.2 A case study of the shortest radius of the small-interference security domain

Numerical simulations are carried out on the two-region system shown in Figure 5.5 with a base power of 100 MVA. In the following, if there is no description of the power, it is the standard per unit value.

The system has seven generators and three loads. The initial power generation is 30.98, the initial load is 30, and the initial damping ratio of the system matrix is 0.2468. From the initial state, the upper limit of the power increase in the most dangerous power increase direction is set to ΔP_{allow}^{max}. If the upper limit of the power is not exceeded, the node injection power increase will not destroy the small-disturbance stability of the system.

Clearly, when the iteration at step W reaches $\Delta P_{allow}^{max} = 0$, we find the shortest radius in the small-disturbance stability domain. For the system shown in Figure 5.5, power growths on nodes 14 and 22 tend to degrade system stability significantly. The power growth step is set as 0.5, while the damping threshold is set as 0.05, and the numerical perturbation step is set as 0.05. After five iterations of calculation, the minimum load margin is 250 MW. That is, no matter what proportion of the added 250 MW load is distributed to each load node and no matter which power generation units bear the added 250 MW load, the system always remains to be stable with small disturbances. The obtained minimum load margin can be used as an important indicator for monitoring and analyzing the small-disturbance stability of the power system.

The location of the "most unfavorable node" for each iteration from the initial state is likely to change. As shown in Table 5.5, the unfavorable nodes for the first iteration are 14 and 22, while at the 5th iteration, power increments on nodes 4 and 25 become more dangerous.

Table 5.5: Each iteration of the system of small-disturbance stability of the nodes with the most unfavorable load and most unfavorable power generation.

Number of iterations	The most unfavorable load node number	The most unfavorable power generation unit node number
1	14	22
2	14	22
3	4	22
4	4	22
5	4	25

5.3.4.3 An example of the shortest radius of the transient security domain

For the system shown in Figure 5.5, a fault happens on the bus bar 24. At $t = 0$ s, a three-phase short circuit occurs, which is then is cleared at $t = 0.1$ s. The stability margin of the system is 0.2468 in the initial state. In the initial state, the increase of the injected power on nodes 14 and 22 is the most dangerous, which can quickly deteriorate the transient stability of the system. Take the load and generation power increase step size as 0.5, and the numerical perturbation step size as 0.05. After five iterations of calculation, the minimum load margin is 200 MW, which means that no matter what proportion of the added 200 MW load is distributed to each load node and which power generation units bear the added 200 MW load, the system will maintain transient stability. The obtained minimum load margin can be used as an important indicator for monitoring and analyzing the transient stability of the system. The most dangerous power increase position for the transient stability of the system at each iteration is shown in Table 5.6.

Table 5.6: The most unfavorable transient load and power generation unit node numbers.

Number of iterations	The most unfavorable load node number	The most unfavorable power generation unit node number
1	14	22
2	14	22
3	4	22
4	4	22
5	4	25

5.4 Coordinated control performance indicator of interconnected power system

Many new problems need to be studied to improve the dynamics and dynamic behavior of interconnected power systems. In European countries and the United States, the interconnected power systems are often governed by different main power companies optimize. These interconnected power systems often face challenges in coordinated controls and operations, whose performances should be quantitatively evaluated by suitable indicators.

At present, there are already A1, A2, CPS1, CPS2, and other standard indicators for active power control of interconnected power grids, which will be briefly introduced in Section 5.4.1. However, there are no corresponding mature indicators for the control of reactive power and voltage in interconnected power grids. For this reason, using the basic idea of area control error (ACE) indicator inactive power control for reference, this book proposes a voltage zone control performance indicator that solves this problem well. See section 5.4.2 for details.

5.4.1 Interconnected power grid active control performance indicator

5.4.1.1 Regional control deviation indicator

The ACE is defined as the sum of the line power deviations and the frequency differences of the control areas. The formula is as follows:

$$\begin{aligned} \mathrm{ACE}_i &= \Delta P_T - 10B_i\Delta f \\ &= (P_a - P_s) - 10B_i(f_a - f_s) \end{aligned} \tag{5.41}$$

where P_s is the active power exchange planned, which is positive if it flows out of the regional power system; P_a is the actual exchanging active power between the area i and other areas; B_i is the frequency response coefficient of regional power system i, whose unit can be MW/0.1 Hz. According to the power system frequency response characteristics, B_i should always be negative; f_a is the actual frequency of the power system, and f_s is the rated frequency of the power system.

5.4.1.2 A1 and A2

The North American Electric Reliability Council, now known as the North American Electric Reliability Corporation (NERC) launched A1 and A2 in 1973 [43]. The A1 standard requires that the area control error (ACE) be zero again within 10 min after the last zero crossing; that is, the time interval between the two zeroes cannot exceed 10 min.

The A2 standard requires that the average of the ACE indicator in the control area must be less than the given value every 10 min. The given value is given by

$$L_d = (0.025)\Delta L + 5\ (\text{MW}) \tag{5.42}$$

where ΔL in the formula can be given by either of the following algorithms:

(1) The maximum value (increase or decrease) of the amount of electricity change in the control zone during the peak loading period of winter or summer in two adjacent hours

(2) The average value of the change in the amount of electricity in any 10 hours of the control zone in a year.

In general, L_d of each control area is adjusted once a year.

By applying the A1 and A2 standards, the ACEs in each control area can be kept close to zero, thus maintaining the balance among the electricity load, the planned and actual active power exchanges. However, the A1 and A2 standards do have some shortcomings as follows:

(1) The main purpose of controlling the ACE is to ensure the quality of the system frequency, but the A1 and A2 standards do not reflect such requirements.

(2) The A1 standard requires that the ACE should always cross zero, thereby increasing the unnecessary regulations of corresponding generators in some cases.

(3) Since the 10 min average of the ACE is strictly controlled, it is difficult for one regional power system to provide enough power support without modifying the exchange plan when an accident occurs in another regional power system.

For these reasons, NERC began to modify performance indicators of coordinated frequency control in interconnected power systems in 1983. Finally, after years of exploration, in 1996, CPS1 and CPS2 were announced as new and standard indicators.

5.4.1.3 Introduction of the CPS1 and CPS2 standards

Now, it is widely accepted that for the automatic generation control(AGC), basic performance criteria are set to keep CPS1 ≥ 100% and CPS2 ≥ 90% within a certain period. The following is a brief introduction of the CPS1 and CPS2 indicators.

(1) CPS1

Unlike the A1 standard, the CPS1 does not imply that the ACE reaches zero within a certain period, thus reducing many unnecessary regulations. At the same time, the CPS1 raises explicit requirements on the frequency quality of the system.

The CPS1 can be calculated using the following formulas:

$$\text{CF} = \text{AVG}_{\text{Period}}\left[\left(\frac{\text{ACE}_i}{-10B_i}\right)_1 \times \Delta F_1\right] / \varepsilon_1^2 \tag{5.43}$$

where ACE_i is the average value of ACE of the control zone i; B_i is the frequency deviation coefficient; ε_1 is the root mean square value of the frequency deviation of the interconnected power grid in 1 min, and ΔF_1 is the minute average frequency deviation of the interconnected power system NERC gives the control performance indicator as CPS1 evaluated by (5.44),

$$\text{CPS1} = (2 - \text{CF}) \times 100\% \tag{5.44}$$

When the CPS1 of a control zone is greater than 200%, it means that this zone is helpful to the frequency quality of the entire interconnected power grid during this period; when the CPS1 is between 100% and 200%, it indicates that the control behavior of this zone is effective, ensuring that grid frequency and tie-line power control quality do not exceed the allowable range. NERC requires that the CPS1 index should be greater than or equal to 100% for a period.

(2) CPS2

The CPS2 indicator is similar to the A2 indicator, which describes the variation scope of the ACE. The difference is that compared with A2, the CPS2 is designed to relax restrictions on the ACE variation limits, thereby facilitating mutual supports between regions in case of emergency. The formulation of CPS2 is as follows:

$$\text{CPS2} = \left(1 - \frac{\text{VI}}{N - N_u}\right) \times 100\% \tag{5.45}$$

where VI is the sum of the frequency offset values generated within every 10 min counted in a month. It can be calculated by the following formula.

$$\text{VI} = \sum_{i=1}^{N - N\mu} \text{VI}_{10\text{-min}} \tag{5.46}$$

where N is the number of 10 min cycles and $N\mu$ means the number of such cycles that are not to be evaluated during the month.

Additionally, a new operation standard can be established incorporating the CPS2 indicator, which can be presented as follows.

$$\text{VI}_{10\text{-min}} = 0, \text{if } \text{AVG}_{10\text{-min}}(\text{ACE}_i) \le L_{10}$$

$$\text{VI}_{10\text{-min}} = 1, \text{if } \text{AVG}_{10\text{-min}}(\text{ACE}_i) > L_{10}$$

where $\text{AVG}_{10\text{-min}}(\text{ACE}_i)$ represents the average of the i-region frequency control deviation in 10 min and L10 is determined by the following equation:

$$L_{10} = 1.65\varepsilon_{10}\sqrt{(-10B_i)(-10B_s)} \tag{5.47}$$

where the value of ε_{10} is the control target value of the root mean square value of the 10-min average frequency deviation of the interconnected power grid in a given year. B_i is the frequency deviation coefficient, and B_s is the sum of the frequency deviation coefficient of each control zone. NERC assumes that the average of the 10 min ACE should follow a normal distribution with a mean as $\sigma = \varepsilon_{10}\sqrt{(-10B_i)(-10B_s)}$. Then, it can be known that the probability of an acceptable frequency variation in the range of $(-1.65\sigma, 1.65\sigma)$ is 90%, which coincides with the operational constraints of the CPS2 indicator.

5.4.2 Research on the reactive power control performance of an interconnected power grid

As described in section 5.4.1, the area control error and frequency performance indicators can be used for coordinating automatic generation controls. Power system reactive power/voltage control requires a similar indicator. In this regard, this section presents a regional voltage control deviation indicator (VACE).

Since the French Electricity Company (EDF) in 1972 began to propose and implement "secondary voltage control," automatic voltage control (AVC) systems have been widely used in domestic and international power grids to optimize the voltage quality, improve the system safety level, reduce net losses, and reduce dispatcher labor intensity [44, 45]. The currently implemented AVC systems are often operated independently for area power systems only. It assumes that adjacent areas or different levels of power grids are relatively weak in reactive power coupling. Thus, the coordination between AVC systems has not been fully considered. However, with the development of the power system, the number of tie-lines between areas is increased, strengthening active and reactive power exchanges at the same time. Consequently, it is difficult to ensure the weak coupling between reactive power control areas. Failure to carry out proper coordination will cause the following problems: (1) Control oscillation may occur, or even cause oscillation instability; (2) The dynamic reactive power reserve in each area is unreasonable, which will not only cause the power generation unit in some areas to lose the voltage regulation ability, which in turn weakens the voltage stability of the whole power system, and also increases the system network loss.

To solve the above problems, EDF and some scholars proposed the coordinated secondary voltage control (CSVC) in the mid-1980s [45, 46], which aims at optimizing the voltage profile of the whole system by regulating the reactive power output of all generators.

For the multi-level dispatch system in China, Sun Hongbin et al. [47] proposed in 2007: The upper-level power system dispatching center guides the lower-level ones through regulating coordinated variables of voltage controls; the lower-level power grid conducts automatic voltage controls to not only meet objectives of local power systems but also track the set value of the coordinated variable given by the upper-level dispatch center in real-time. The above scheme divides the voltage control responsibilities of different areas and different levels by coordination variables. Thus, the amount of information exchanged during coordinated controls tends to be small and does not involve the specific control logic of local AVCs. Obviously, it can be seen that carefully choosing coordination variables becomes critical for enhancing the efficiency and effectiveness of coordinated voltage controls. Conventionally, the gateway voltage and reactive power (power factor) are taken as the coordination variable. Coordinations using such variables may encounter the following problems: (1) The gate voltage or reactive power is determined by the behavior of multiple interconnected systems, which may not be suitable for assessing the control performance of a single area fairly; (2) The use of these coordination variables cannot avoid the aforementioned issues of control oscillation and unbalanced reactive power reserve.

This book adopts new ideas to solve the above problems: (1) The voltage ACE (VACE) is proposed, which can distinguish between reactive power disturbances in this area or adjacent areas, and measure the influences of local control actions on reactive power flows in adjacent areas; (2) Unless reactive power disturbances occur in this area, local reactive power regulations should be as few as possible, so as to clarify control responsibilities of interconnected area power systems.

5.4.2.1 Sensitivity analysis of tie nodes

Without loss of generality, suppose that an area power system (denoted as I in short) is connected to the external power system (denoted as E in short) through M tie lines, as shown in Figure 5.5.

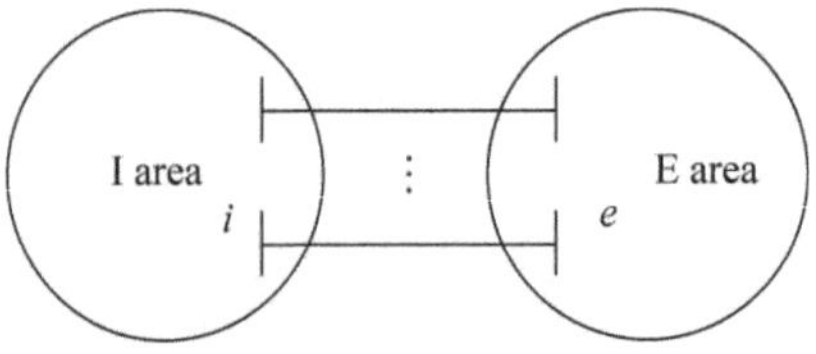

Figure 5.5: Area power system with external connections.

Regardless of the changes in the reactive load and generation in the E area, the I area will be affected by the voltage variations of the tie line nodes. Therefore, it is necessary to analyze the influence of such voltage variations on the voltage of the tie line node i in the I area and the reactive power flow of the tie line. Then, the VACE indicator can be defined accordingly.

First, we examine the effect of the voltage change at node e on the voltage of tie-node i. The influence of the voltage on the reactive power change can be determined by the Q-V iterative equation used in the fast decoupled power flow calculation as shown in (5.48) [48],

$$-\begin{bmatrix} \boldsymbol{B}_D & \boldsymbol{B}_{DG} & \boldsymbol{B}_{Di} & 0 & 0 \\ \boldsymbol{B}_{GD} & \boldsymbol{B}_G & \boldsymbol{B}_{Gi} & 0 & 0 \\ \boldsymbol{B}_{iD} & \boldsymbol{B}_{iG} & \boldsymbol{B}_i & \boldsymbol{B}_{ie} & 0 \\ 0 & 0 & \boldsymbol{B}_{ei} & \boldsymbol{B}_e & \boldsymbol{B}_{eR} \\ 0 & 0 & 0 & \boldsymbol{B}_{Re} & \boldsymbol{B}_R \end{bmatrix} \begin{bmatrix} \Delta\boldsymbol{V}_D \\ \Delta\boldsymbol{V}_G \\ \Delta\boldsymbol{V}_i \\ \Delta\boldsymbol{V}_e \\ \Delta\boldsymbol{V}_R \end{bmatrix} = \begin{bmatrix} \Delta\boldsymbol{Q}_D \\ \Delta\boldsymbol{Q}_G \\ \Delta\boldsymbol{Q}_i \\ \Delta\boldsymbol{Q}_e \\ \Delta\boldsymbol{Q}_R \end{bmatrix} \tag{5.48}$$

where $\boldsymbol{B}$ is the nodal susceptance matrix. The subscript i represents the tie-line node on the side of the I area on behalf of the tie-line node on the side of the E area; the subscript D represents all the PQ nodes in the I area except the tie-line node; G represents all the PV nodes in the I area except the tie-line node and slack node; R stands for all nodes in the E area except the tie line node.

The node with the subscript D in equation (5.48) can be eliminated as follows:

$$-\begin{bmatrix} \widetilde{\boldsymbol{B}}_G & \widetilde{\boldsymbol{B}}_{Gi} & 0 & 0 \\ \widetilde{\boldsymbol{B}}_{iG} & \widetilde{\boldsymbol{B}}_i & \boldsymbol{B}_{ie} & 0 \\ 0 & \boldsymbol{B}_{ei} & \boldsymbol{B}_e & \boldsymbol{B}_{eR} \\ 0 & 0 & \boldsymbol{B}_{Re} & \boldsymbol{B}_R \end{bmatrix} \begin{bmatrix} \Delta\boldsymbol{V}_G \\ \Delta\boldsymbol{V}_i \\ \Delta\boldsymbol{V}_e \\ \Delta\boldsymbol{V}_R \end{bmatrix} = \begin{bmatrix} \Delta\boldsymbol{Q}_G \\ \Delta\boldsymbol{Q}_i \\ \Delta\boldsymbol{Q}_e \\ \Delta\boldsymbol{Q}_R \end{bmatrix} \tag{5.49}$$

where

$$\widetilde{\boldsymbol{B}}_i = \boldsymbol{B}_i - \boldsymbol{B}_{iD}(\boldsymbol{B}_D)^{-1}\boldsymbol{B}_{Di}$$

When a reactive power disturbance occurs in the E region, the reactive power generation and load in the area I can be assumed to remain unchanged, that is,

$$\Delta\boldsymbol{Q}_i = 0 \tag{5.50}$$

$$\Delta\boldsymbol{V}_G = 0 \tag{5.51}$$

Substitute (5.50) and (5.51) into equation (5.49) to obtain the relationship between the voltage changes of node i and e as follows:

$$\Delta\boldsymbol{V}_i = -\widetilde{\boldsymbol{B}_i^{-1}}\boldsymbol{B}_{ie}\Delta\boldsymbol{V}_e \tag{5.52}$$

The sensitivity matrix of the voltage of node i on the voltage of node e is $\boldsymbol{C}_{ie} = -\widetilde{\boldsymbol{B}}_i^{-1}\boldsymbol{B}_{ie}$.

Suppose that there are multiple connection lines in parallel connecting two boundary nodes, the susceptance matrix of boundary nodes of area I and E is

$$\boldsymbol{B}_{\text{ie}} = -\begin{bmatrix} b_{l1} & \cdots & 0 \\ \vdots & \ddots & \vdots \\ 0 & \cdots & b_{lm} \end{bmatrix}$$

where $b_{l1}, b_{l2}, \ldots, b_{lm}$ is the susceptances of tie lines. Now, make

$$\boldsymbol{b}_l = -\boldsymbol{B}_{ie}$$

Let's examine the influence of the voltage change of the tie line node e in area E on the reactive power of tie-line node i in area I. The reactive power flowing into the node i in area I through the tie line can be obtained by the following formula

$$\widetilde{Q_{lk}} = (b_{lk} + b_{ck})V_{ik}^2 + (g_{lk}\sin\theta_{ikek} - b_{lk}\cos\theta_{ikek})V_{ik}V_{ek}, \quad k = 1, 2, \ldots, m \tag{5.53}$$

where b_{ck} is the 1/2 charging susceptance of each tie line; g_{lk} is the conductance of each tie line; $\theta_{i_k e_k}$ is the vector phase angle difference between the two ends of the kth line; and V_{ik} and V_{ek} are the i-node and e-node voltage values, respectively.

As the angle difference between the ends of the line is generally small enough, it can be approximated that $\sin\theta_{ik,ek} \approx 0$ and $\cos\theta_{ik,ek} \approx 1$; then, equation (5.53) can be simplified as

$$\widetilde{Q_{ik}} = (b_{lk} + b_{ck})V_{ik}^2 - b_{lk}V_{ik}V_{ek}, \quad i = 1, 2, \ldots, m \tag{5.54}$$

Equation (5.54) is approximately linearized at a rated operating point $V_{ik}^0 = V_{ek}^0 = 1$ pu or in a sufficiently small neighborhood of the other specified operating point I area. Then, we have

$$\Delta\widetilde{\boldsymbol{Q}_{ik}} = 2(\boldsymbol{b}_{lk} + \boldsymbol{b}_{ck})\Delta\boldsymbol{V}_{ik} - b_{lk}\Delta\boldsymbol{V}_{ek} - \boldsymbol{b}_{lk}\boldsymbol{V}_{ik}, \tag{5.55}$$

By substituting (5.52) into (5.55), the relationship between $\Delta\widetilde{Q_{Qi}}$ and ΔV_e can be obtained as

$$\Delta\widetilde{\boldsymbol{Q}}_i = (\boldsymbol{b}_l\boldsymbol{C}_{ie} - \boldsymbol{b}_l + 2\boldsymbol{b}_c\boldsymbol{C}_{ie})\Delta\boldsymbol{V}_e \tag{5.56}$$

5.4.2.2 Voltage area control error indicator, VACE

The area control error (ACE) [49] is defined in AGC as follows:

$$\text{ACE} = (P_T - P_{\text{ref}}) + B(f - f_{\text{ref}}) \tag{5.57}$$

When B is equal to the natural frequency characteristic coefficient, the ACE can distinguish the location of a frequency disturbance. Each area is responsible for maintaining local power balance and providing temporary supports to neighboring areas in case of an emergency. Similar to the ACE, a linear combination of the boundary node voltage and the reactive power of the tie line is given in (5.58), which defines the VACE indicator

$$\textbf{VACE}_I = \left(\widetilde{\boldsymbol{Q}}_i - \widetilde{\boldsymbol{Q}}_{\text{iref}}\right) + \boldsymbol{E}_I(\boldsymbol{V}_i - \boldsymbol{V}_{\text{iref}}) \tag{5.58}$$

where $\widetilde{\boldsymbol{Q}}_{\text{iref}}$ and $\boldsymbol{V}_{\text{iref}}$ are the set value of the node voltage and flow-in reactive power of the local side of the tie line.

$\mathbf{VACE}_l$ can also be written in incremental form; that is,

$$\boldsymbol{VACE}_I = \Delta\widetilde{\boldsymbol{Q}}_i + \boldsymbol{E}_I \Delta \boldsymbol{V}_i \tag{5.59}$$

where

$$\Delta\widetilde{\boldsymbol{Q}}_i = \widetilde{\boldsymbol{Q}}_i - \widetilde{\boldsymbol{Q}}_{\text{iref}}.$$

The expression of E_{I} is derived by substituting (5.56) into (5.59):

$$\begin{aligned}\boldsymbol{VACE}_{\text{I}} &= (\boldsymbol{b}_l \boldsymbol{C}_{ie} - \boldsymbol{b}_l + 2\boldsymbol{b}_c \boldsymbol{C}_{ie})\Delta \boldsymbol{V}_e + \boldsymbol{E}_{\text{l}} \boldsymbol{C}_{ie} \Delta \boldsymbol{V}_e \\ &= (\boldsymbol{b}_l \boldsymbol{C}_{ie} - \boldsymbol{b}_l + 2\boldsymbol{b}_c \boldsymbol{C}_{ie} + \boldsymbol{E}_l \boldsymbol{C}_{ie})\Delta \boldsymbol{V}_e \end{aligned} \tag{5.60}$$

Choosing an appropriate E_I makes: For any external reactive power disturbance, the value of VACE in this area does not change, that is

$$\mathbf{VACE}_I = \mathbf{0} \tag{5.61}$$

From equation (5.60), we can obtain

$$\boldsymbol{b}_l \boldsymbol{C}_{ie} - \boldsymbol{b}_l + 2\boldsymbol{b}_c \boldsymbol{C}_{ie} + \boldsymbol{E}_I \boldsymbol{C}_{ie} = \mathbf{0} \tag{5.62}$$

where

$$\boldsymbol{E}_I = \boldsymbol{b}_l \boldsymbol{C}_{ie}{}^{-1} - \boldsymbol{b}_l - 2\boldsymbol{b}_c \tag{5.63}$$

By substituting (5.52) into (5.63), we have

$$\boldsymbol{E}_{\text{I}} = \widetilde{\boldsymbol{B}}_{\text{i}} - \boldsymbol{b}_l - 2\boldsymbol{b}_c \tag{5.64}$$

The meaning of selecting parameters according to formula (5-64) can ensure that no matter how the reactive power of the external area E fluctuates, as long as the reactive power distribution of the I area remains unchanged (the Q of the PQ node or the V of the node does not change), then the VACE of area I remains unchanged.

The **VACE** indicator has the following properties:

(1) The VACE indicator is positive, indicating that the voltage in this area is low compared with the reference value or too much reactive power is absorbed from the outside, while the reactive power compensation is insufficient; the VACE indicator is negative, indicating that the voltage in this area is too high or the reactive power is injected to the outside massively, while reactive power compensation is excessing;
(2) If the VACE indicator is equal to 0 after control, it means that the influence of local reactive power changes on adjacent areas has been eliminated. Ideally, when the adjacent areas are controlled to maintain their respective VACE indicators near 0, these areas are regulating their reactive power flow independently while the boundary node voltage and reactive power of the tile lines reach the reference set value.
(3) VACE has the same unit as reactive power, and the calculation is simple and easy to assess.

5.4.2.3 Coordinated AVC based on the VACE index

With the **VACE** indicator, the area AVC system takes charge of not only regulating the system voltage profile but also maintaining the VACE indicator within a range close to 0 to coordinate other areas' control actions.

First, the control sensitivity of reactive power outputs of generators in area I to the VACE indicator is derived. It is not hard to derive the following relationship.

$$\Delta\widetilde{Q}_i = \left(\boldsymbol{b}_l - \boldsymbol{b}_l\widetilde{\boldsymbol{B}_e^{-1}}\boldsymbol{b}_l + 2\boldsymbol{b}_c\right)\Delta V_i \tag{5.65}$$

From (5.59) we can have

$$\Delta \boldsymbol{VACE}_I = \Delta\widetilde{\boldsymbol{Q}}_i + \boldsymbol{E}_I\Delta\boldsymbol{V}_i$$

Substituting (5.71) and (5.64) into the above formula makes the following equations available:

$$\begin{aligned}\Delta \boldsymbol{VACE}_I &= \left(\boldsymbol{b}_l - \boldsymbol{b}_l\widetilde{\boldsymbol{B}_e^{-1}}\boldsymbol{b}_l + 2\boldsymbol{b}_c\right)\Delta\boldsymbol{V}_i + \left(\widetilde{\boldsymbol{B}}_i - \boldsymbol{b}_l - 2\boldsymbol{b}_c\right)\Delta\boldsymbol{V}_i \\ &= \left(\widetilde{\boldsymbol{B}}_i - \boldsymbol{b}_l\widetilde{\boldsymbol{B}_e^{-1}}b_l\right)\Delta\boldsymbol{V}_i\end{aligned}$$

Then, we can define

$$\mathbf{C}_{VACE\mathrm{i}} = \widetilde{\boldsymbol{B}}_i - \boldsymbol{b}_l\widetilde{\boldsymbol{B}_e^{-1}}\boldsymbol{b}_l \tag{5.66}$$

which is the sensitivity of the voltage of boundary node i on the VACE indicator.

Further, with the sensitivity matrix $\boldsymbol{C}_{iG}$, the sensitivity of the VACE indicator to each AVC generator can be obtained as follows:

$$\boldsymbol{C}_{\mathrm{VACE}G} = \boldsymbol{C}_{\mathrm{VACE}i}\boldsymbol{C}_{iG} \tag{5.67}$$

Therefore, to control the VACE in a range close to zero, the AVC system in each area does not need to change the original optimization objective. Only the following constraints need to be added to the optimal control formulation of local AVC systems.

$$-\boldsymbol{\delta} \le \mathbf{VACE} + \boldsymbol{C}_{\mathrm{VACE}G}\Delta\boldsymbol{V}_G \le \boldsymbol{\delta}, \quad \boldsymbol{\delta} > \mathbf{0} \tag{5.68}$$

where δ is the maximum deviation allowed by the VACE indicator. δ should take the appropriate value according to the actual situation. If the value is too small, the new optimization problem may be unsolvable; if the value is too large, the goal of coordination will not be achieved.

5.4.2.4 Case study

The case study uses the New England 39-node system (see Figure 5.6). The whole system can be divided into two interconnected systems connected by three tile-lines

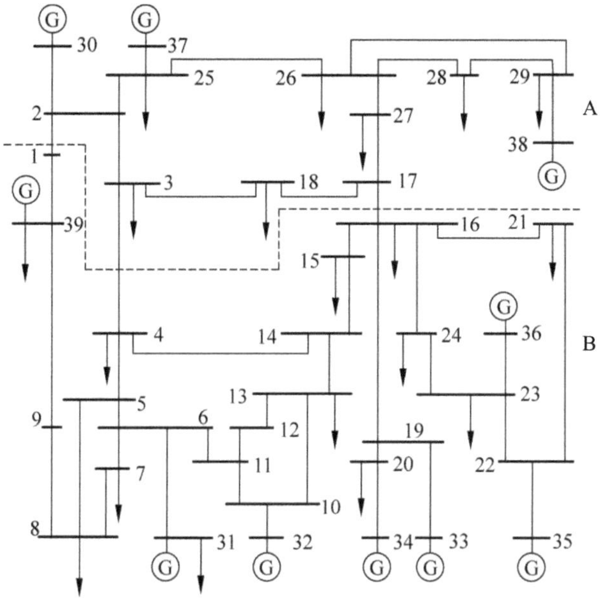

Figure 5.6: IEEE 39 bus system divided into two regions.

such as 1-2, 3-4, and 16-17. As shown in Figure 5.8, the upper area (above the dotted line) is abbreviated as A, and the lower area is abbreviated as B.

First, we verify the effectiveness of the **VACE** indicator. The reactive load of PQ nodes 7, 15, and 24 is gradually increased by 10 Mvar (0.1 pu) in the 10 time steps. Then, the reactive power disturbance happens in area B in the following 10 time steps while voltages of PV node 37 and 38 increase 1kV (0.01 p.u.).

As shown in Figure 5.7, when the reactive load in the B zone gradually increased, the **VACE** value exhibited a linear increase. It means that the reactive power supply in area B is insufficient, and too much reactive power is absorbed from the adjacent area. At the same time, since there is no reactive power disturbance in area A, its VACE indicator remains constant. When the setting value of the generator terminal voltage in the A area gradually increases, the VACE value of the A area decreases linearly, which means that the area reactive power supply is so excessive that too much reactive power is injected into the adjacent area. At the same time, there is no reactive power disturbance in the B area, while the VACE indicator of area B remains constant. Moreover, it can be seen from Figure 5.8 that the VACE indicator of the 1-2 tie-line in area B has always remained unchanged. This is because node 39 (PV node) connected to node 1 blocks the influence of the reactive load disturbance on it.

The following is a comparison of the control performances with and without coordination using the VACE indicator.

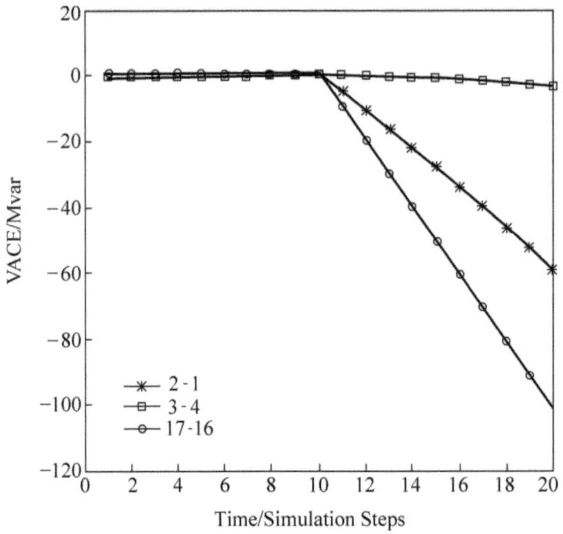

Figure 5.7: Dynamics of the VACE index of area A.

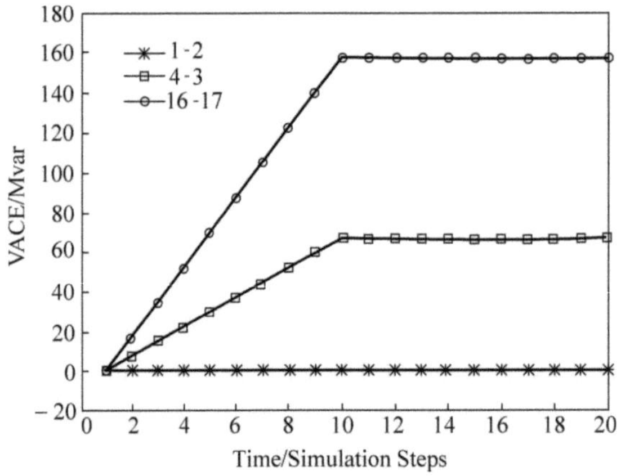

Figure 5.8: Dynamics of the VACE index of area B.

Assuming that both areas are equipped with an AVC system, a conventional secondary voltage control algorithm is used. The key nodes selected in area A are nodes 3 and 27, and the generators on nodes 30 and 37 are regulated by the AVC system. The key nodes in area B are nodes 7 and 13, and the generators one the node 39, 32, 33, 34, 35, and 36 are controlled.

Scenario A: For each time step of the simulation, two AVC controllers of areas A and B solve their local optimal control strategies the control commands and simultaneously execute corresponding controls in the next time step.

Scenario B: Coordinated by the VACE indicators, corresponding constraints are added to optimal control formulations of the AVC system of both areas. Then, procedures in scenario A are carried out properly.

At the 20th time step, the reactive load of node 15 in area B suddenly increases by 100 Mvar (1 p.u.), and the two area AVC systems take action at the same time.

Representing by the key node voltage profiles, the control performances of the voltage controllers in the two areas before and after the coordination are shown in Figure 5.9 and Figure 5.10. It can be seen that with coordination, the key node voltage is much closer to the set value than the one with uncoordinated controls. Meanwhile, the voltage fluctuation during control processes is relatively small when the coordination is conducted by using the VACE indicator.

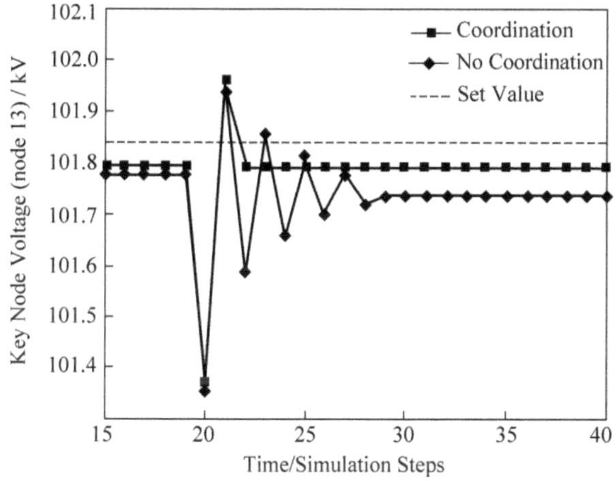

Figure 5.9: Comparisons of the voltage of node 13 with and without coordinated controls.

In the absence of coordination, the generators in the two areas have more obvious control oscillations, which disappear after nearly ten adjustments. When coordination is used, the control oscillations are significantly suppressed in 3 adjustments. Meanwhile, the voltage variations of key nodes are eliminated rapidly, while fewer adjustments of reactive power generations are required. Since the VACE-based coordination restricts influences of local reactive power generations on the adjacent areas, unnecessary overshoot and repeated adjustments are reduced.

The generator reactive power margins of both areas are shown in Figure 5.11. Without coordination, area reactive power margins are unbalanced, where the reactive power generation in area A is close to its lower limit, and the reactive power generation in area B is close to its upper limit. After using the VACE indicator to coordinate,

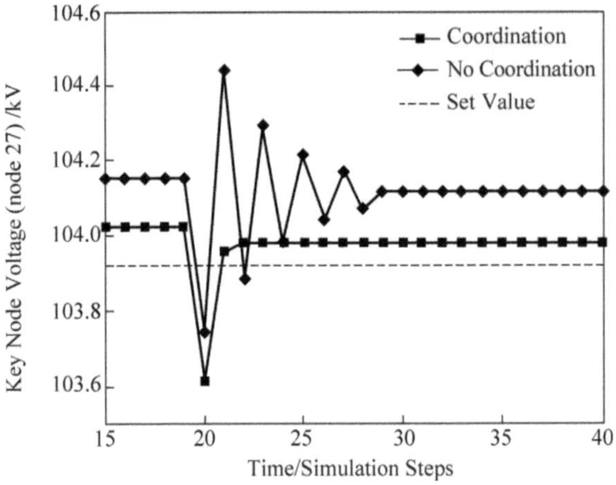

Figure 5.10: Comparisons of the voltage of node 27 with and without coordinated controls.

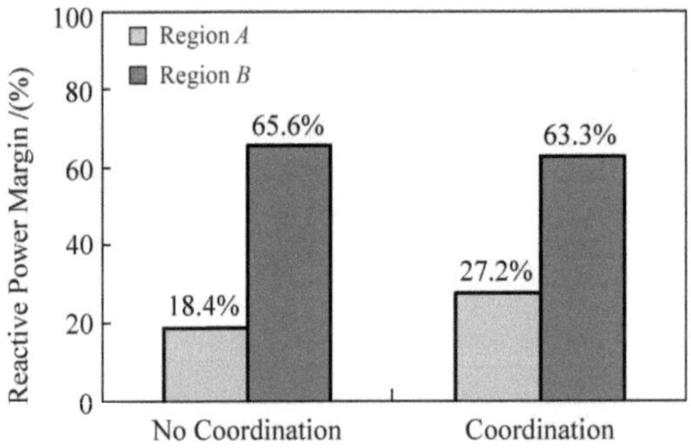

Figure 5.11: Comparisons of reactive power margins with and without coordinated controls.

the reactive power margins of the two areas tend to be more balanced. Consequently, the active power loss of the entire system is reduced, and the loss is reduced by 0.1 MW, which is equivalent to 0.2% of the overall power generation.

It can be seen from this example that the coordination of area AVC using the VACE indicator can enhance the performance of system-level voltage and reactive control, that is, optimizing reactive power reserve, suppressing control oscillations, and reducing network losses. If there is a superior dispatch center optimizing and issuing optimized VACE indicator reference for each tie-line, reasonable reactive power support between area power systems can be achieved as well as the balanced reserve of area reactive power generation capabilities.

In summary, the proposed VACE indicator has a clear physical meaning and can quantitatively describe the mutual influence between voltage control areas.

Using VACE, the control performance, and fairness of the coordinated voltage control of interconnected power systems can be enhanced remarkably.

5.5 Summary

Defining operation performance indicators is indispensable to realize the "multi-index self-approximate optimization" of the smart power system. This chapter introduces the standard operation performance index system of smart power systems. First, the design method of the power system operation performance indicators is introduced. From the perspectives of ideas, compositions, and computation processes, various power system operation performance indicators, as well as their interrelationships, are described comprehensively. Furthermore, this chapter proposes several novel safety indicators and coordinated control performance indicators of interconnected power systems. Theoretical analyses and numerical simulations are also presented to validate the proposed indicators.

It should be pointed out that to achieve "multi-index self-approximate optimization," the SEMS of the smart power system needs continuously evaluate operation performance indicators and identify unacceptable system states according to over-threshold events of these indicators. When such events happen, it is necessary to execute necessary control actions to readjust system states and eliminate these events eventually. It can be seen that only by combining with the hybrid control theory, the operation standard index system of the smart power system can play an important role.

Chapter 6
Event analysis and processing technology

6.1 Introduction

The goal of a smart power system is to achieve multi-index optimality-approximating operation of a power system. Chapter 5 introduces how to use the standard indicator system of power system operation to quantify multiple indices, and this chapter introduces how to use event analysis and processing technology to achieve multi-index optimality-approximating operation of power systems.

The power hybrid control theory (PHCT), introduced in Chapter 2, is the theoretical basis of the optimality-approximating operation of power systems. According to the PHCT, power systems will be under optimality-approximating operation conditions as long as the power systems are running in an event-free state. In contrast, once these systems are in an unsatisfied state (i.e., an event state), auto control measures should be taken to make the power systems return to the event-free state. Therefore, the optimality-approximating process of power systems is exactly the process of event analysis and processing. Thus, two key technological problems need to be solved:

(1) Reliability of data. Achieving the optimality-approximating operation of power systems implies achieving closed-loop automatic control of power systems. Closed-loop automatic control requires data to have high credibility. Taking control measures based on insufficiently credible data may lead to unknown consequences. As is well known, the processing unit to improve data credibility is called "state estimation." A weighted least squares (WLS) algorithm is usually used in traditional state estimation systems. Since the results are susceptible to bad data, the algorithm cannot meet the requirements of closed-loop automatic control. For solving this problem, a new state estimation algorithm based on new ideas is proposed in Section 6.2.
(2) Desirable convergence, reliability, and rapidity are crucial for analyzing and calculating power systems. The optimal power flow (OPF) calculation is the most frequently used calculation method in event analysis and processing. The convergence of the OPF calculation is generally poor. Hence, the computing speed is slow, thus making the OPF calculation difficult to apply in closed-loop automatic control. For this problem, a new OPF algorithm is proposed in Section 6.3.

https://doi.org/10.1515/9783110448825-006

6.2 Advanced State Estimation (ASE) algorithm

Power system state estimation using a WLS estimation algorithm was first proposed by F. C. Schweppe [50, 51]. For data that follow the normal distribution, WLS estimation is the minimum variance unbiased estimator, and its mathematical model and calculation are simple. A disadvantage is that, in practical situations, the distribution of the measurement errors is always far from the normal distribution, thus leading to a loss of the excellent characteristics of WLS estimation. Moreover, the results of WLS estimation are not immune to large deviations in individual measurements.

To overcome the shortcomings of WLS estimation, the robust estimation theory, which is defined as an estimation method that can resist model deviations and observation disturbances, has been proposed [52]. The goals of robust estimation include the following: estimation results are optimal or near-optimal under the assumed model; when the deviation between the actual model and the assumed model is small, the effect on the estimation is small; and when the deviation is large, the effect on the estimation is not catastrophic. The robust estimation methods for power systems mainly include maximum likelihood (M) estimation [53], generalized M (GM) estimation [54], and high breakdown contamination rate estimation [55] (the breakdown contamination rate is the ratio of the largest proportion of the number of measurements with arbitrarily large gross errors that can be handled by the estimation method to the total number of measurements). M estimation, namely, GM estimation, includes minimum absolute value and weighted minimum absolute value estimation [56], quasi-likelihood (QL) and QC estimation [57], and so on. For weakening the adverse impact of a bad leverage measurement, GM estimation was proposed [54]. By increasing the spatial information of the system structure to improve the robust structure capability of the estimation, the breakdown contamination rate is improved. Since M estimation and GM estimation cannot handle an abundance of bad data, and the breakdown contamination rate is still low, Rousseeuw and Leroy proposed a high breakdown contamination rate estimation method [55], including minimum median squares estimation and least trimmed square estimation. High breakdown contamination rate estimation improves robustness by selecting the measurement information but does not take full advantage of all the valid measurement information. In addition, the computational efficiency is very low, and the computing time increases exponentially as the size of the system increases.

The above methods are based basically on the concept of the residual (the difference between an estimated value and a measured value), and their basic idea is to make the estimated values an exact fit to the measured values by minimizing the weighted sum of residuals. However, the measured values are not the true values and also include uncertainties. A measurement point with the minimum residual indicates that its estimated value is close to the measured value instead of the true value. Moreover, because of the existence of measurement errors, the estimated value may be far from the true value if a precise fitting between the estimated value and the measured value is pursued excessively. Therefore, an estimation method that aims to minimize a certain norm of residuals is vulnerable to bad data. Additionally, the method is not robust.

The main reason for the above difficulties is the limitation of measurement residuals. Any information about the true values of the measurement points cannot be inferred from the residuals, thus making a quantitative description of the reliability of the measurement results impossible. Therefore, after years of research, international metrology scholars decided to use measurement uncertainty to characterize the reliability of measurement results. Measurement uncertainty is a parameter associated with measurement results and is used to characterize the dispersion of values reasonably attributed to measurement values.

Based on measurement uncertainty, a new kind of power system state estimation method, named advanced state estimation (ASE), is proposed in this section. Instead of pursuing a precise fit between estimated values and measured values, this method maximizes the estimated values located in the measurement uncertainty interval of the selected measured values (i.e., this method takes the maximum normal measurement point rate as the indicator) by using the information of measurement uncertainty. The method is robust, and the estimated results are not vulnerable to bad data. Furthermore, a load flow solution that satisfies various constraints and is closer to the actual situation can be obtained using this method. Additionally, no bad data validations, observability validations, or manual setting of measurement point weights is needed; so, debugging and maintenance work will be greatly reduced. In a concrete realization, the proposed method can exhibit good convergence and a fast computational speed using a modern interior-point algorithm [58].

The specific arrangement of the rest of this section is as follows: In Section 6.2.1, the concept of measurement uncertainty is introduced, and the differences between measurement uncertainty and measurement error are noted. In Section 6.2.2, the main idea of the ASE method is presented, and in Section 6.2.3, the ASE method is elaborated. In Section 6.2.4, the characteristics of the ASE method are summarized. Finally, in Section 6.2.5, the ASE method is verified through a simulation case study, and the results are compared with those of a WLS method and a QC method.

6.2.1 Measurement uncertainty

In 1993, the Guide to the Expression of Uncertainty in Measurement (GUM) was issued by seven international organizations, including the International Standards Organization (ISO). Measurement uncertainty, reflecting the possible error distribution, can be approximatively explained as the error limits under a certain confidence level. Measurement uncertainty can be divided into two categories: standard uncertainty and expanded uncertainty.

Standard uncertainty is the uncertainty of a measurement, expressed as one standard deviation. The measurement value is Z_i; if the true value is $\bar{Z}_i$ and the uncertainty under the confidence probability p is u, the probability that the true value is located in the interval $[Z_i - u, \dot{Z}_i + u]$ is p, which is given as follows:

$$P(|\overline{Z}_i - Z_i| \leq u) = p \tag{6.1}$$

When the uncertainty follows the normal distribution, p is 68.3%.

The expanded uncertainty, U, is a value defining the interval of measurement results and is used to express the confidence range of the measurement results. This value can be obtained by multiplying the combined standard uncertainty u with the coverage factor k as follows:

$$U = ku, \quad k \in \mathbf{N} \tag{6.2}$$

When k is 3 and the corresponding uncertainty follows the normal distribution, the confidence level P is 99.7%.

Conceptually, measurement uncertainty differs significantly from measurement error. Measurement error is defined as the difference between a measurement result and the true value. Because the true value is unknown, the error is an ideal concept that generally cannot be actually known and is difficult to quantify and handle. The measurement uncertainty u, which denotes the dispersion of the measurement values, shows how little is known about the measurement. This uncertainty is an interval that takes the estimated value as the standard and can be obtained by analysis and assessment; additionally, this uncertainty can be quantified and is easy to handle.

If the information regarding measurement uncertainty cannot be obtained or if someone wants to obtain the estimation state with the largest qualified rate during the practical application, the expanded measurement uncertainty can be replaced by the defined value of the qualified rate α_i based on the industry standard. The principle of selection of α_i is as follows:

$$\alpha_i = \begin{cases} 0.02 \times z_{ibase}, & \text{node } i \text{ is the voltage measurement point} \\ 0.02 \times z_{ibase}, & \text{node } i \text{ is the active power measurement point} \\ 0.03 \times z_{ibase}, & \text{node } i \text{ is the reactive power measurement point} \end{cases} \tag{6.3}$$

In (6.3), the reference z_{ibase} is defined as follows:

(1) For the measurement of voltage amplitude, z_{ibase} is set to 1.2 times the rated voltage. For example, when the rated voltage is 500 kV, z_{ibase} is 600 kV; when the rated voltage is 330 kV, z_{ibase} is 396 kV; and when the rated voltage is 220 kV, z_{ibase} is 264 kV.

(2) For the measurement of the power of a generator, z_{ibase} is set to the apparent power of the generator.

(3) For the measurement of power flow, z_{ibase} is set to 1,083 MV · A when the voltage of the line is 500 kV; z_{ibase} is set to 686 MV · A when the voltage is 330 kV; z_{ibase} is set to 305 MV · A when the voltage is 220 kV; z_{ibase} is set to 114 MV · A when the voltage is 110 kV; and z_{ibase} is set to 69.7 MV · A when the voltage is 66 kV.

6.2.2 Main idea of ASE

In brief, the main idea of ASE is to maximize the normal measurement point rate directly. The discussion is below.

The measurement equation of measurement points in power systems can be expressed as follows:

$$Z_i = h_i(x) + e_i, \quad i = 1, 2, \cdots, m \tag{6.4}$$

where $\boldsymbol{X}$ is the system state; Z_i and $h_i(x)$ are the measurement value and measurement function, respectively, of the measurement point i in the system; e_i is the measurement error of point i, and m is the number of measurement points in the system.

The absolute value of the measurement residual of point i can be obtained from (6.4) as follows:

$$|e_i| = |Z_i - h_i(x)| \tag{6.5}$$

In practical applications, normal measurement points and abnormal measurement points can be defined according to the relationship between the measurement residual and the measurement uncertainty.

Definition 6.1 For measurement point i, the expanded measurement uncertainty under the confidence probability p is Ui, and $\boldsymbol{X}$ is the system state. Z_i and $h_i(x)$ denote the measurement value and the measurement function, respectively. If (6.6) holds true, measurement point i is called a normal measurement point; otherwise, it is called an abnormal measurement point.

$$|h_i(\mathbf{X}) - Z_i| \le U_i \tag{6.6}$$

ASE adopts the following new ideas, which differ from those of the existing state estimation methods:

(1) First, it is assumed that the measurement points containing bad data are only a small fraction of all points. Apparently, this assumption meets the actual situation of the power system.
(2) The information of measurement uncertainty is introduced. The true measurement values fall in the interval determined by the measurement uncertainty, with a probability of approximately 1. For any given estimation state X_ε, if the measurement point i is a normal point, according to (6.6), the measurement

value Z_i of point i is considered to agree with the estimation; else, Z_i is considered to disagree with the estimation.

(3) Instead of pursuing the minimum error between the estimated value and measurement value, ASE considers a state in which most measurement points agree with the estimation (namely, most measurement points fall within the selected measurement uncertainty interval) as the most reasonable estimation state and takes it as the final estimation result.

6.2.3 Introduction of the ASE algorithm

6.2.3.1 Evaluation function of measurement points

Definition 6.2 Assuming that $\boldsymbol{X}$ is the state estimation result, the relative deviation *di* of $\boldsymbol{X}$ at measurement point i is defined as follows:

$$d_i = (h_i(X) - Z_i)/U_i \tag{6.7}$$

where Z_i is the measurement value of point i, $h_i(\cdot)$ is the measurement function of point i, and U_i is the expanded measurement uncertainty of point i under the confidence probability p.

Definition 6.3 The measurement evaluation function, $g(d_i)$, is defined as

$$g(d_i) = \begin{cases} 0, & |d_i| \le 1 \\ 1, & |d_i| > 1 \end{cases} \tag{6.8}$$

The shape of $g(d_i)$ is shown in Figure 6.1. When the absolute value of the relative deviation of the estimation state $|d_i| \le 1$, the measurement point i is a normal point, and the value of $g(d_i)$ is zero. When $|d_i| > 1$, the point i is an abnormal point, and the value of $g(d_i)$ is one.

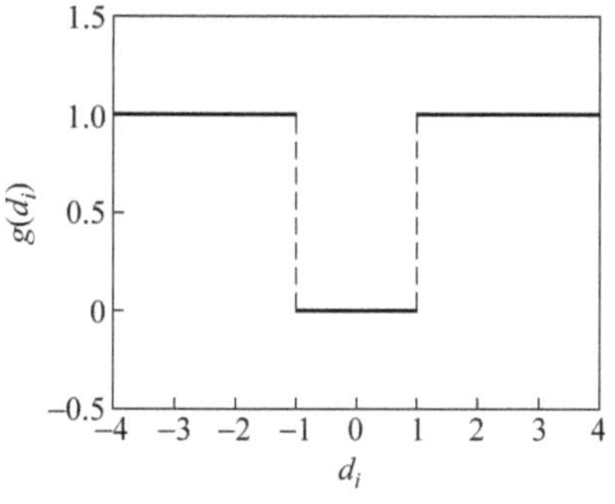

Figure 6.1: The measurement evaluation function $g(d_i)$.

Definition 6.4 If N is the number of normal measurement points in a certain power system and m is the total number of measurement points, N/m is called the normal measurement point rate and is denoted as N_m.

Definition 6.5 $\sum_{i=1}^{m} g(d_i)$ is defined as the evaluation indicator of the normal measurement point rate, Nm, under-estimation state X.

According to (6.7) and (6.8), an estimation state that minimizes the evaluation indicator of the normal measurement point rate maximizes the normal measurement point rate N_m. Thus, the state estimation problem of seeking the maximum N_m can be transformed into an estimation problem of seeking the minimum evaluation indicator of the normal measurement point rate.

In practical applications, abnormal measurement points usually need to be divided into wrong points and suspicious points. A point is considered a suspicious point when $|d_i|$ is slightly greater than 1. When $|d_i|$ is significantly greater than 1, the point is considered an abnormal point.

Definition 6.6 λ denotes a given constant greater than 1. If measurement point i meets the condition of $1 < |d_i| < \lambda$, i is considered a suspicious point; otherwise, it is considered an abnormal point.

To ensure that the measurement evaluation function $g(d_i)$ is close to zero when $|d_i| \le 1$, $g(d_i)$ is close to one when $|d_i| \ge \lambda$ and $g(d_i)$ is continuously differentiable everywhere,

The curve of $g(d_i)$ shown in Figure 6.1 is smoothed for practical applications. Therefore, a practical measurement evaluation function $f(d_i)$ is selected as follows:

$$f(d_i) = \delta(d_i) + \delta(-d_i) \tag{6.9}$$

where $\delta(d_i)$ is actually the sigmoid function and is given by

$$\delta(d_i) = \frac{1}{1 + e^{-k(ad_i + b)}} \tag{6.10}$$

To make $f(\pm\lambda) \approx 1$ and $f(\pm 1) \approx 0$, the parameters k, a, and b are set according to the following rules:

$$a = \frac{2}{\lambda - 1} \tag{6.11}$$

$$b = -1 - \frac{2}{\lambda - 1} \tag{6.12}$$

The practical measurement evaluation function $f(d_i)$ ($k = 3$ and $\lambda = 2$) is shown in Figure 6.2.

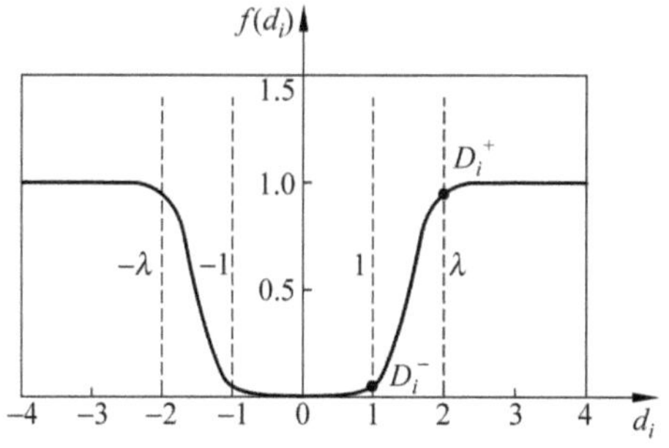

Figure 6.2: The practical measurement evaluation function $f(d_i)$ ($k = 3$, $\lambda = 2$).

The values of $f(\pm 1)$ and $f(\pm\lambda)$, with different values of k, are listed in Table 6.1. Table 6.1 shows that the greater k is, the closer to zero (one) $f(\pm 1)$ ($f(\pm\lambda)$) is. However, when k is too large, it adversely affects the convergence of optimization problems whose objective functions are to minimize $\sum_i f(d_i)$.

Table 6.1: Values of the function $f(d_i)$.

k	$f(\pm 1)$	$f(\pm\lambda)$
1	0.2869	0.7335
1.5	0.1849	0.8177
2	0.1195	0.8808
2.5	0.0759	0.9241
3	0.0474	0.9526
3.5	0.0293	0.9707
4	0.0180	0.9820

6.2.3.2 Mathematical model

Since $\sum_{i=1}^{m} f(d_i)$ is approximately equal to the number of abnormal measurement m points, seeking a system state with many normal points is almost equivalent to seeking a system state where $\sum_{i=1}^{m} f(d_i)$ is small. Thus, the ASE model is developed as follows:

$$\min_x \sum_{i=1}^{m} f(d_i) \tag{6.13}$$

The real system state must meet power flow constraints and the upper and lower limits of operational constraints. Then, the following state estimation model can be obtained:

$$\begin{aligned} &\min_{X} \sum_{i=1}^{m} f(d_i) \\ &s.t.\quad d_i = (h_i(X) - Z_i)/U_i, \forall i = 1, 2, \cdots, m \\ &\qquad\ \ \mathbf{g}(X) = \mathbf{0} \\ &\qquad\ \ \mathbf{l}(X) \le \mathbf{0} \end{aligned} \tag{6.14}$$

where $g(x) = 0$ represents power flow constraints and $l(x) \le 0$ represents actual physical constraints, such as the upper and lower limits of generator outputs.

6.2.3.3 Solution algorithm

By inserting (6.7) into the optimization indicator $\sum_{i=1}^{m} f(d_i)$ in (6.14), the following state estimation model can be obtained:

$$\begin{aligned} &\min_{X} \sum_{i=1}^{m} f((h_i(X) - Z_i)/U_i) \\ &s.t.\quad \mathbf{g}(X) = \mathbf{0} \\ &\qquad\ \ \mathbf{l}(X) \le \mathbf{0} \end{aligned} \tag{6.15}$$

This optimization problem can be solved in many ways. A modern interior-point algorithm has many advantages, such as good convergence properties, high computing speed, and an insignificant increase in the calculation, with an increase in system size. Therefore, here, we choose a modern interior point algorithm to solve the problem. However, (6.15) is actually a nonconvex and nonlinear programming problem. The modern interior-point algorithm can guarantee convergence; however, the algorithm cannot guarantee that the global optimal solution can always be obtained. Furthermore, the estimation results depend on the initial values. Hence, the initial values should be chosen properly. The analysis is given as follows.

As shown in Figure 6.2, the $f(d_i)$ curve has four turning points symmetrically distributed on both sides of the vertical axis that divide the horizontal axis into three intervals: $(-\lambda, -1)$, $(-1, 1)$, and $(1, \lambda)$. To facilitate discussion, the two turning points on the right part of the curve are labeled D_i^- and D_i^+, from left to right (see Figure 6.2).

Under a certain system state X, the location of D_i^+ changes with the parameter λ. When the value of λ decreases, D_i^+ approaches D_i^-. Thus, when the value of λ is too small, the distance $|D_i^+, D_i^-|$, between D_i^+ and D_i^-, will also be too short. Thus, for most measurement points, when $|d_i| > D_i^+$ or $|d_i| < D_i^-$, the evaluation ability of $f(d_i)$ is strong. However, the nonconvexity of $f(d_i)$ is also strong, thus possibly leading to the problem that the optimization has difficulty in converging. Conversely, if λ is too large, the nonconvexity and the evaluation ability of $f(d_i)$ become weak, thus reducing the ability of the function to significantly identify bad data.

In specific applications, it is recommended to solve the problem in two steps. First, choose a large number as the initial value of λ (such as $\lambda = 5.0$) for the optimization calculation. Not only is a large value beneficial in using the modern interior point method to solve the problem, but the ability of the measurement evaluation function to identify bad data is also maintained. Second, take the result of the first step as the initial value, and decrease the value of λ (for example, $\lambda = 2.0$). Solve the optimization problem again. Repeat this step, and a satisfactory estimation result can be obtained.

6.2.4 Features of the ASE algorithm

Compared with previous state estimation methods, the method proposed in this book has the following features:

(1) Measurement uncertainty information is introduced into the state estimation model, which may fully consider the uncertainty of the measurement itself. The new method does not pursue an accurate data fit between the estimation value and the measurement value. For a measurement point that is abnormal, in this method, regardless of how far the estimation value is from the measurement value, the value of the objective function is one (see Figures 6.1 and 6.2). Therefore, the estimation results are robust.
(2) The method can automatically detect bad data during state estimation. Abnormal measurement points obtained during estimation can be regarded as detected bad data.
(3) In practical applications, if the measurement uncertainty of measurement points cannot be known, it can be replaced by the qualified range of measurement points defined by industry standards. Thus, the obtained estimation state has the largest qualified rate.
(4) The obtained state estimation result is a power flow solution that satisfies the real constraints of all types of devices. Therefore, this method is more realistic. Additionally, no bad data validations or manual setting of the measurement point weights is needed in the calculation; so, the debugging and maintenance work of the software is simple. The solution method is a modern interior-point algorithm that exhibits good convergence and high computational speed.

6.2.5 Case study

The introduced state estimation method is applied to the IEEE 30-bus and 118-bus systems, and the results are compared with those of the traditional WLS and QC methods.

6.2.5.1 Bad data validation capability verification

In this example, 2% Gaussian noise is added to the power flow results, and three times the standard deviation is taken as the expanded uncertainty.

For the IEEE 30-bus system, the measurement settings are the same as those in [60]. Six bad data points are included; they are listed in Table 6.2. As shown in Table 6.2, the power flow measurements of lines 1-2 and the power injection measurement of bus 1 are relevant bad data. The power flow measurements of lines 24-25 and the power injection measurement of bus 29 are irrelevant bad data [60].

Table 6.2: Bad data in the IEEE 30-bus system.

Data type	Normal measurement point (pu)	Bad data (pu)
P_{1-2}	1.7786	0.0000
Q_{1-2}	−0.2562	0.2200
P_1	2.5825	0.0000
P_{24-25}	−0.0127	0.2000
P_{29}	−0.0244	−0.1200
Q_{29}	−0.0089	−0.1200

With the ASE method, six bad data points are detected and identified as abnormal measurement points, while other data are regarded as normal points. The detected abnormal points are listed in Table 6.3.

Table 6.3: Abnormal data detected by the ASE method (IEEE 30-bus system).

Data type	True value	Measurement value	Results of ASE
P_{1-2}	1.7795	0.0000	1.7891
Q_{1-2}	−0.2578	0.2200	−0.2576
P_1	2.6103	0.0000	2.6257
P_{24-25}	−0.0123	0.2000	−0.0107
P_{29}	−0.0240	−0.1200	−0.0225
Q_{29}	−0.0090	−0.1200	−0.0088

Similarly, the ASE method is applied to the IEEE 118-bus system; four relevant bad data points are set; they are listed in Table 6.4.

Table 6.4: Bad data in the IEEE 118-bus system.

Data type	Normal measurement data (pu)	Bad data (pu)
P_{1-2}	−0.1297	−0.0620
Q_{1-2}	−0.1314	0.0000
P_1	−0.5169	−0.2550
Q_1	−0.3074	−0.1500

With the ASE method, four bad data points are detected and identified as abnormal measurement points, while the others are considered as normal points. The detected abnormal measurement points are listed in Table 6.5.

Table 6.5: Abnormal measurement points detected by ASE (IEEE 118-bus system).

Data type	True value	Measurement value	Results of ASE
P_{1-2}	−0.1240	−0.0620	−0.1237
Q_{1-2}	−0.1265	0.0000	−0.1306
P_1	−0.5147	−0.2550	−0.5115
Q_1	−0.3027	−0.1500	−0.3022

6.2.5.2 Comparison with the WLS and QC methods

The ASE method is compared with the WLS and QC methods mentioned in [57] in the above IEEE 30-bus and 118-bus cases. All three methods use a flat start. The parameters used in the QC method are the same as those used in [57]. The results are shown in Tables 6.6 and 6.7. n_{bad} denotes the number of bad data points detected by the estimator. Both wrong and suspicious measurement points are included for the ASE method. The deviations of the voltage amplitude and phase angle, namely, E_V and E_θ, respectively, are provided by [57], and their definitions are:

$$E_V = \frac{1}{n}\sum_{i=1}^{n}(V_i - \hat{V}_i)^2$$

$$E_\theta = \frac{1}{n}\sum_{i=1}^{n}(\theta_i - \hat{\theta}_i)^2$$

where n is the bus number; V_i and θ_i are the true values of the voltage amplitude and phase angle, respectively; and $\hat{V}_i$ and $\hat{\theta}_i$ are the estimation values of the voltage

Table 6.6: Comparison with WLS and QC methods (IEEE 30-bus case with 6 bad data points).

Methods	n_{bad}	E_V	E_θ
ASE	6(6bad)	1.3447×10^{-5}	0.0030
WLS	—	6.5184×10^{-4}	14.4368
QC	18(3bad + 15good)	9.7288×10^{-5}	30.2966

amplitude and phase angle, respectively. The units of the voltage amplitude are per unit, and the units of the phase angle are degrees.

Tables 6.6 and 6.7 show that the ASE method can precisely detect bad data in both test cases, and the indices of voltage amplitudes and phase angles are better than those used with the WLS method and the QC method. Especially for the IEEE 30-bus case, the QC method can detect only three bad data points, and 15 good data points are considered as bad data points (see Table 6.6). Due to the effect of bad data, the deviations of voltage amplitudes and phase angles in the WLS method and the QC method are large. The ASE method can precisely detect the six bad data points, and the deviations of voltage amplitudes and phase angles are small. In the IEEE 118-bus case, the QC method incorrectly identifies one datum as a bad datum, while the ASE method detects all four bad data points with no errors (see Table 6.7).

Table 6.7: Comparison with WLS and QC methods (IEEE 118-bus case with 4 bad data points).

Methods	n_{bad}	E_V	E_θ
ASE	4(4bad)	1.4202×10^{-9}	1.7200×10^{-4}
WLS	—	2.0480×10^{-7}	3.5293×10^{-4}
QC	5(4bad + 1good)	7.9885×10^{-8}	2.2089×10^{-4}

6.3 An OPF algorithm based on the constraint transformation technology

To introduce OPF into power system control, one of the problems that must be solved is the challenge in obtaining a power flow solution that is better than the current state when no feasible solution exists due to tight constraints. Consequently, a practical OPF algorithm in which constraints in power systems are divided into hard constraints and soft constraints is proposed in this section. Using the constraint transformation technology in the calculation ensures that the algorithm can still converge to a better solution than that at the current state, even if the constraints are too strict.

6.3.1 OPF model

The objective function of the OPF model is to minimize the generation cost; that is,

$$\min f(\boldsymbol{P}) = \sum_{i \in S_G} (a_i P_{G_i}^2 + b_i P_{G_i} + c_i). \tag{6.16}$$

where $f(P)$ represents the total generation cost function of the power system; P_{G_i} represents the output of the ith generation unit; SG represents the set of all generation units; and a_i, b_i, and c_i represent the coefficients of the output polynomial of the ith generation unit.

The OPF model should satisfy the following constraints:

Power Flow Equation Constraints

$$\begin{cases} P_{G_i} - P_{D_i} - \sum\limits_{j=1}^{n} V_i V_j \left| Y_{ij} \right| \cos(\theta_i - \theta_j - \delta_{ij}) = 0 \\ Q_{R_i} - Q_{D_i} - \sum\limits_{j=1}^{n} V_i V_j \left| Y_{ij} \right| \sin(\theta_i - \theta_j - \delta_{ij}) = 0 \end{cases} \tag{6.17}$$

Operation Inequality Constraints

$$\begin{cases} \underline{V}_i \le V_i \le \bar{V}_i, & i \in S_N \\ \underline{P}_{G_i} \le P_{G_i} \le \bar{P}_{G_i}, & i \in S_G \\ \underline{Q}_{R_i} \le Q_{R_i} \le \bar{Q}_{R_i}, & i \in S_R \\ \underline{I}_{ij} \le I_{ij} \le \bar{I}_{ij}, & i,j \in S_L \end{cases} \tag{6.18}$$

where V_i represents the voltage amplitude of the ith node and $\underline{V}_i$, $\bar{V}_i$ represents the lower and upper limits, respectively, of the node; P_{G_i} denotes the active power of the ith generation unit; $\underline{P}_{G_i}$, $\bar{P}_{G_i}$ is the lower and upper limits, respectively, of the active power of the ith generation unit; Q_{R_i} represents the reactive power of the ith reactive power resource; $\underline{Q}_{R_i}$, $\bar{Q}_{R_i}$ represent the lower and upper limits, respectively, of the reactive power of the ith reactive power resource; I_{ij} is the current of branch $i-j$; $\underline{I}_{ij}$,$\bar{I}_{ij}$ represent the lower and upper limits, respectively, of the current of branch $i-j$; and SG, SN, SR, and SL represent the sets of generation units, nodes, reactive power resources, and branches, respectively.

In the above optimization problem, the power flow constraints and generation output constraints together called the hard constraints, should be satisfied. The other constraints, such as the nodal voltage constraints and power limit constraints of branches, are called soft constraints since a small violation will not lead to serious consequences for a power system during peak load time periods.

Therefore, the OPF model can be simplified as follows:

$$\begin{aligned} &\min f(X) \\ s.t.\ \ &g(X) = 0 \\ &\underline{h}_H \le h_H(X) \le \bar{h}_H \\ &\underline{h}_S \le h_S(X) \le \bar{h}_S \end{aligned} \tag{6.19}$$

where $f(x)$ is the generation cost function of the state vector of the whole power system; $h_H(x)$, which corresponds to hard constraints, represents the constraints of the output of generation units; $h_S(x)$, which corresponds to soft constraints, represents the nodal voltage constraints and the power limit constraints of branches; and $\underline{h}$ and $\bar{h}$ represent the lower and upper limits, respectively of h.

By introducing the slack variables of hard constraints, s_l^H and s_u^H, and the slack variables of soft constraints, s_l^S, and s_u^S, into the optimization problem, (6.19) can be reformulated as follows:

$$\begin{aligned} &\min f(X) \\ s.t.\ \ &g(X) = 0 \\ &h_H(X) - s_l^H - \underline{h}_H = 0 \quad s_l^H \ge 0 \\ &\bar{h}_H - h_H(X) - s_u^H = 0 \quad s_u^H \ge 0 \\ &h_S(X) - s_l^S - \underline{h}_S = 0 \quad s_l^S \ge 0 \\ &\bar{h}_S - h_S(X) - s_u^S = 0 \quad s_u^S \ge 0 \end{aligned} \tag{6.20}$$

A new objective function is built using the logarithmic barrier function

$$\varphi_\mu(x) = f(X) - \mu\left(\sum_{i\in H} (\ln(s_l^H(i)) + \ln(s_u^H(i))) + \sum_{j\in S} (\ln(s_l^S(j)) + \ln(s_u^S(j)))\right) \tag{6.21}$$

where μ is a parameter greater than zero and declines toward zero during iterations. According to (6.21), it is easy to know that when $\mu \to 0$, $\varphi_\mu(x) \to f(x)$.

By introducing the barrier function, the original optimization problem is transformed into the following equations:

$$\begin{aligned} &\min \varphi_\mu(X) \\ s.t.\ \ &g(X) = 0 \\ &h_H(X) - s_l^H - \underline{h}_H = 0 \\ &\bar{h}_H - h_H(X) - s_u^H = 0 \\ &h_S(X) - s_l^S - \underline{h}_S = 0 \\ &\bar{h}_S - h_S(X) - s_u^S = 0 \end{aligned} \tag{6.22}$$

.

The Lagrange function is defined as follows:

$$\begin{aligned}L(x,y,z_l,z_u) = {} & \varphi_\mu(X) - y^T g(X)\\ & - z_l^H(h_H(X) - s_l^H - \underline{h}_H)\\ & - z_u^H(\bar{h}_H - h_H(X) - s_u^H)\\ & - z_l^S(h_S(X) - s_l^S - \underline{h}_S)\\ & - z_u^S(\bar{h}_S - h_S(X) - s_u^S)\end{aligned} \tag{6.23}$$

where $x, s_l^H, s_u^H, s_l^S, s_u^S$ are original variables and $y, z_l^H, z_u^H, z_l^S, z_u^S$ are dual variables.

According to the first-order Kavrush-Kuhn-Tucker (KKT) conditions, the following equation is obtained:

$$\begin{bmatrix} \nabla f(x) - \nabla g(X)^T y - \nabla h_H(X)^T(z_l^H - z_u^H) \\ - \nabla h_S(X)^T(z_l^S - z_u^S) \\ g(X) \\ h_H(X) - s_l^H - \underline{h}_H \\ \bar{h}_H - h_H(X) - s_u^H \\ h_S(X) - s_l^S - \underline{h}_S \\ \bar{h}_S - h_S(X) - s_u^S \\ Z_l^H S_l^H e - \mu e \\ Z_u^H S_u^H e - \mu e \\ Z_l^S S_l^S e - \mu e \\ Z_u^S S_u^S e - \mu e \end{bmatrix} = \mathbf{0} \tag{6.24}$$

Equation (6.24) is abbreviated as $K(x,\mu) = 0$. In (6.24), e is a column vector with a length equal to the dimension of the corresponding constraint. The elements of e are all one: $Z_l^H := diag(z_l^H)$, $Z_u^H := diag(z_u^H)$, $Z_l^S := diag(z_l^S)$,$Z_u^S := diag(z_u^S)$, $S_l^H := diag(s_l^H)$, $S_u^H := diag(s_u^H)$, $S_l^S := diag(s_l^S)$, and $S_u^S := diag(s_u^S)$. The notation $diag(\cdot)$ denotes a diagonal matrix composed of the elements of the vector in the brackets.

According to (6.24), when $\mu \to 0$, (6.24) will be the solution to the optimization problem (6.19). When (6.24) is solved iteratively, the search direction, dx, can be calculated by Newton's method. The steps of the original and dual variables are set to be different. α_p denotes the steps of the original variables, and α_d denotes the steps of the dual variables. α_p and α_d are represented by the following equations:

$$\begin{cases} \alpha_p = \min(1, min(s + \alpha_p ds \geq 0)) \\ \alpha_d = \min(1, min(z + \alpha_d dz \geq 0)) \end{cases} \tag{6.25}$$

where s and z represent the original variables and the corresponding dual variables, respectively; and ds and dz represent the search directions of the original and the dual variables, respectively. Thus, the new operation state $x(k+1)$ and the slack variables can be obtained by iteration as follows:

$$\begin{cases} x(k+1) = x(k) + \alpha_p dx \\ s_l^S(k+1) = s_l^S(k) + \alpha_p ds_l^S \\ s_u^S(k+1) = s_u^S(k) + \alpha_p ds_u^S \\ s_l^H(k+1) = s_l^H(k) + \alpha_p ds_l^H \\ s_u^H(k+1) = s_u^H(k) + \alpha_p ds_u^H \end{cases} \tag{6.26}$$

In the actual (practical) iterative process, the original variables are updated by using the above method. However, this method does not guarantee that the updated solutions will lead to a smaller objective function or fulfill the constraints better. Normally, we shorten the steps and recalculate if this happens. However, in many cases, this approach is ineffective. Therefore, a better choice is to correct the feasible direction.

When correcting the feasible direction, the primary objective is to minimize the extent to which the current operation point exceeds the soft constraints, and the optimization model is given as follows:

$$\begin{aligned} &\min f(x) = \sum_{s^S<0} \|s^S\|_1 + \frac{\xi}{2}\|x - x_R\|_2^2 \\ s.t.\quad & g(x) = 0 \\ & h_H(x) - s_l^H - \underline{h}_H = 0 \\ & \bar{h}_H - h_H(x) - s_u^H = 0 \end{aligned} \tag{6.27}$$

where ξ is a given constant, $s^S = [s_l^S, s_\mu^S]^T$ is the slack variable vector of the soft constraints, and x_R is the variable x, before the correction along the feasible search direction. To avoid a large increase in the objective function after the correction, we add $\frac{\xi}{2}\|x - x_R\|_2^2$ to the objective function so that the distance between x and x_R (x before the correction) is not too large.

The model described in (6.27) is similar to that in (6.20), and the solution method is the same. Therefore, it is not repeated here.

6.3.2 Workflow of the OPF algorithm

The above problem can be solved with the interior point OPF method based on a filter set [61, 62]. The solution steps are summarized as follows.

Step 1: Initialize the original and dual variables

Set the initial values of the original and dual variables of the optimization problem shown in (6.22), that is, $x(0), s_l^s(0), s_u^s(0), s_l^H(0), s_u^H(0)$ and $y(0), z_l^s(0), z_h^s(0), z_l^H(0), z_h^H(0)$. Calculate the objective function of the initial operation point, θ, and the extent to which the constraints are exceeded, φ. φ is calculated by using the following equation:

$$\varphi = \left\| \begin{matrix} g(\mathbf{X}) \\ h_H(\mathbf{X}) - s_l^H - \underline{h}_H \\ \overline{h}_H - h_H(\mathbf{X}) - s_u^H \\ h_S(\mathbf{X}) - s_l^s - \underline{h}_S \\ \overline{h}_s - h_s(\mathbf{X}) - S_u^s \end{matrix} \right\|_1 \tag{6.28}$$

The notation $\|\bullet\|_1$ denotes the 1-norm, and the definition is

$$\varphi = \left\| \begin{matrix} \varphi_1 \\ \varphi_2 \\ \varphi_3 \\ \varphi_4 \\ \varphi_5 \end{matrix} \right\|_1 = \sum_{i=1}^{5} |\varphi_i|$$

Let $F_0 := \{(\theta, \varphi) \in R^2\}$. F_k denotes the set of the pair of θ and φ (the filter set) in the kth iteration.

Step 2: Judge the convergence by using the KKT conditions

If $\|K(x, 0)\|_\infty < \varepsilon_{tol}$, the optimal solution is obtained. Thus, the calculation stops and the results are output. Else, the iterative process continues. ε_{tol} represents the tolerance defined by the user.

Step 3: Update μ_j

If $\|K(x, \mu_j)\|_\infty < \kappa_\varepsilon \mu_j$ (κ_ε is a constant greater than zero), then update μ according to the following equation:

$$\mu_{j+1} = \max\left(\frac{\varepsilon_{tol}}{10}, 0.2^* \mu_j\right)$$

Step 4: Solve the problem by iteration

Calculate (6.24) using the Newton's method and obtain the corrected search direction. The step is determined by (6.25). Calculate the corrected variables $x(k+1)$, $x(k+1), s_l^s(k+1), s_u^s(k+1), s_u^H(k+1), s_u^H(k+1)$ and the corresponding θ, φ by using (6.26).

If $\theta, \varphi \in F_k$, then go to Step 5; otherwise, accept this state and let

$$F_{k+1} := F_k \cup \left\{ \begin{array}{l} (\theta, \varphi) \in R^2 : \theta \geq (1 - \gamma_\theta)\theta(x_k) \\ \text{and} \quad \varphi \geq \varphi_{\mu_j}(x_k) - \gamma_\varphi \theta(x_k) \end{array} \right\}$$

where γ_θ and γ_φ are constants, $\gamma_\theta \in (0,1)$ and $\gamma_\varphi \in (0,1)$.

Return to Step 2.

Step 5: Correct the feasible search direction

Solve (6.27). Correct the feasible search direction and return to Step 2.

6.3.3 Case studies

The interior point OPF program based on the filter set is written using MATLAB 7. The test systems include the IEEE 9-bus system, IEEE 30-bus system, IEEE 57-bus system, IEEE 118-bus system, IEEE 300-bus system, Shanghai 216-bus system (SHH216), and the northeast 542-bus system (NE 542). The program runs on a computer with a Pentium IV 3.0 GHz central processing unit (CPU) and 1 GB memory. The operating system is Windows XP.

The basic parameters of the test systems are listed in Table 6.8.

Table 6.8: Statistics of the test systems.

Test system	Bus	Generator	Line	Variable	Equation	Inequation
IEEE 9	9	3	9	24	18	15
IEEE 30	30	6	41	72	60	42
IEEE 57	57	7	80	128	114	71
IEEE 118	118	54	186	344	236	226
SHH 216	216	52	325	536	432	320
IEEE 300	300	69	411	738	600	438
NE 542	542	137	737	1,358	1,084	816

In the case studies, $\varepsilon_{tol} = 10^{-6}$. The maximum number of iterations $k_{\max} = 50$.

The test results with a flat start and hot start are listed in Tables 6.9 and 6.10, respectively.

In the case studies, the proposed method is compared with the OPF method provided by the Power System Analysis Toolbox (PSAT). PSAT is a MATLAB toolbox for comprehensive power system analysis. PSAT is powerful, and its computational

Table 6.9: Calculation results for a flat start.

Test system	Method proposed in this book		PSAT	
	Iterations	Computing time/s	Iterations	Computing time/s
IEEE 9	10	0.36	10	0.31
IEEE 30	8	0.31	7	0.26
IEEE 57	8	0.39	10	0.55
IEEE 118	14	0.82	13	2.82
SHH 216	11	1.09	50+	
IEEE 300	12	1.30	17	47.1
NE 542	15 + 21	8.92	50+	

Table 6.10: Calculation results for a hot start.

Test system	Method proposed in this book		PSAT	
	Iterations	Computing time/s	Iterations	Computing time/s
IEEE 9	7	0.26	7	0.20
IEEE 30	8	0.36	7	0.24
IEEE 57	10	0.55	10	0.53
IEEE 118	11	0.67	13	2.93
SHH 216	9	1.00	15	4.23
IEEE 300	12	1.30	41	97.11
NE 542	20	4.52	50+	

speed is fast. Mehrotra's predictor-corrector interior point method is used in PSAT. The computing processes of Hessian matrices are optimized in version 2.0.0-b1.8 to further improve the OPF calculation speed. The convergence conditions are set to be the same for comparison. In all case studies, as long as the method converges, the OPF result of the proposed method is the same as the result calculated by PSAT. Due to space limitations, the calculation results are not listed here.

Figure 6.3 presents the relationship between the number of buses and the elapsed iteration time. It can be seen that there is an approximately linear relationship between the number of buses and the elapsed iteration time.

By analyzing the comparison results shown in Tables 6.9 and 6.10, the following conclusions can be obtained:

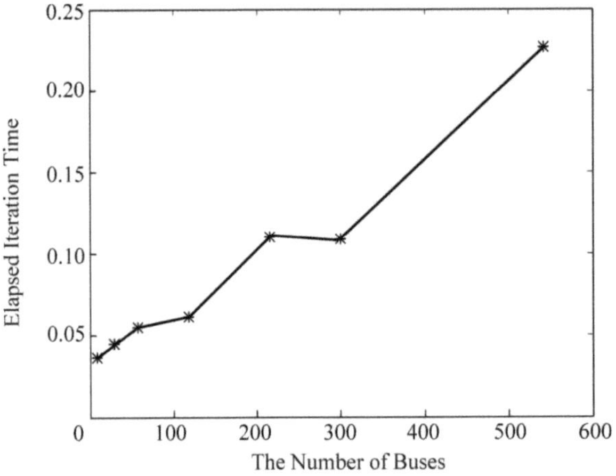

Figure 6.3: Relationship between the number of buses and the elapsed iteration time.

(1) Calculation speed

Regardless of whether a flat start or hot start is used, the number of iterations of the method proposed in this book decreases more significantly than that of the PSAT method, with increasing test system size. Correspondingly, the calculation speed of the proposed method improves more significantly.

(2) Sensitivity to dependence on initial conditions

The number of iterations and the elapsed time in PSAT differ greatly between flat and hot starts, thus indicating that the method used in PSAT has greater sensitivity to dependence on initial conditions. The method proposed in this book does not have this problem.

(3) Convergence

Overall, both methods exhibit very good convergence for a hot start. The PSAT method does not converge for the Shanghai 216-bus system and the northeast 542-node system for a flat start. The method proposed in this book converges in nine iterations for the Shanghai 216-bus system. When the proposed method is used to test the northeast 542-node system, the feasible direction correction mode starts after 15 iterations. After another 15 iterations, the method converges to a feasible solution. Furthermore, the method converges to the final optimal solution after another seven iterations. When using a hot start, the method used in PSAT does not converge for the northeast 542-node system, while the method proposed in this book converges for all test systems. The test results indicate that the convergence and practicality of the method proposed in this book are better than those of the method used in PSAT.

6.4 Summary

In this chapter, key technologies for event processing and analysis in smart power systems are introduced. The ASE method can provide accurate, comprehensive, and timely information about the states of power systems (the state of the power system is called the real state $\mathbf{X}_{\mathrm{TS}}(t)$) for comprehensive state assessment and analysis of various operation indices. Instead of minimizing the difference between the estimated value and the measurement value, the ASE method takes maximizing the normal measurement point rate as the objective, which is the key difference between the ASE method and the traditional methods. The proposed OPF algorithm is effective and can be used to calculate global optimization control commands online to achieve real-time closed-loop control of power systems.

Chapter 7
Smart power system visualization

7.1 Introduction

For a smart power system, the optimal control adopted should not rely solely on machine intelligence but on a combination of machine intelligence and the artificial intelligence of system operators.

To this end, the smart power system discussed in this book has an inherent advantage. The theoretical basis (hybrid control theory) involves an event-driven kernel, which is highly consistent with the daily operating behavior of system operators based on events. Thus, both event-driven kernels and system operators adopt the same control concept.

The combination of machine intelligence and artificial intelligence also relies on excellent visualization. Excellent visualization systems should be able to provide the most critical information in the most vivid way to the people who demand it most, in due time, so that operators can always know the operating status of a system, where they should intervene, and how they should intervene, to achieve a strong combination of machine intelligence and artificial intelligence.

For the abovementioned requirements, smart power system visualization is largely different from the current power system visualization in content and implementation, which is summarized as follows:

(1) Smart power system visualization is an active research field. Current visualization technology mainly refers to the operating state visualization of the power systems. Smart power system visualization includes not only the visualization of system operation status but also that of the operation indices, events, control commands, operation instructions, instruction execution effects, etc. The details are discussed in Section 7.2.
(2) Because the smart power system visualization of content and the effects thereof are greatly enhanced, graph-generating and graph-drawing methods must also be developed accordingly.
(3) Smart power system visualization places more emphasis on the automatic generation of topological graphics, which is a new requirement, inevitably caused by the highly enriched visualization content. In the current visualization technology, basic graphics are manually drawn and maintained using graphic editors. This method not only creates a large working burden and easily introduces errors but can also only generate static graphics and cannot dynamically generate new graphics based on the user's demand. Moreover, graphical information exchange among different application systems is also very difficult because different systems usually need to maintain different graphic systems. The automatic generation of topological graphics can not only compensate for the

https://doi.org/10.1515/9783110448825-007

abovementioned deficiencies but also improve visual effects. The specific details are described in Section 7.3.

(4) Smart power system visualization has placed stringent requirements on the speed of graphic rendering, as a real-time dynamic display of the system operating status is required so that "the operators can always know the operating status of the system." The drawing speed of the existing graphic technology cannot meet this requirement for complex graphics. Taking the voltage contour map as an example, based on the common regular grid methods, a workstation with a central processing unit (CPU) of 2.0 GHz frequency and a memory of 2 GB requires 1 ~ 3 s to complete a graph with a resolution of 800 × 800. Section 7.4 introduces a novel fast graphic rendering algorithm based on the algorithm of grid merging voltage contour interpolation and graphics generation; this algorithm provides a new way to solve this issue. The method is proposed for high-voltage contours, but the underlying principles can also be applied to other similar situations.

7.2 Smart power system visualization content

Visualization in scientific computing (ViSC), hereafter referred to as visualization, is defined as the use of the principles and methods of computer graphics to translate large-scale data from science and engineering calculations into an intuitive form (i.e., graphics and images).

In current studies, the visualization of power system operation mainly refers to operation status visualization, including the visualization of topological graphics and operation data. Indeed, the visualization of power system operation should also include monitoring visualization; that is, operation state index visualization, event visualization, control command visualization, operation instruction visualization, instruction execution visualization, etc. This research field is still relatively small and should focus on further development.

7.2.1 Operation state visualization

As previously mentioned, operation state visualization can be divided into two parts: topology visualization and operational data visualization, which are described as follows.

7.2.1.1 Topological graphic visualization

A power system is the physical manifestation of a complex network. Therefore, the topological graphic visualization of a power system is the premise and foundation for achieving the visualization of the entire operation of a power system. The topological graphics of power systems can be divided into two categories: single-line diagrams and plant wiring diagrams, which are shown in Figures 7.1 and 7.2, respectively.

(1) A single-line diagram shows the physical connection between power plants and substations. The representation mainly includes power plants, substations, and lines. In actual diagram drawings, line segments are commonly used to represent transmission lines, and dots or squares are used to represent substations. A single-line diagram is sometimes superimposed on a background map with geographic information, as shown in Figure 7.1. This method can truly and intuitively indicate the geographical distribution of a grid structure. Partial magnification can be used to resolve densely distributed local elements.

(2) An electrical wiring diagram is used to represent the physical connections inside power plants and substations. The representation mainly includes transformers, breakers, switches, lines, reactances, and compensation devices. These objects are generally represented by standard patterns in an actual diagram drawing.

Figure 7.1: Example of a single-line diagram.

7.2.1.2 Operation data visualization

There are two main types of operation data in power systems: bus-type data and line-type data. The two different data types require different visualization expressions.

The easiest way to visualize bus-type data is to label the information near the bus location. For example, a rectangle filled with blue and green colors placed in the vicinity of a bus can be used to represent the bus voltage value between the maximum and minimum limits. The amplitude and phase angle of the bus voltage can also be represented by a histogram and a pie chart near the bus.

Another way to visualize bus-type data is to extend the data of several buses defined within a finite range to the entire region using the spatial interpolation method, and then use the visualization method of a 2D scalar field (e.g., color maps,

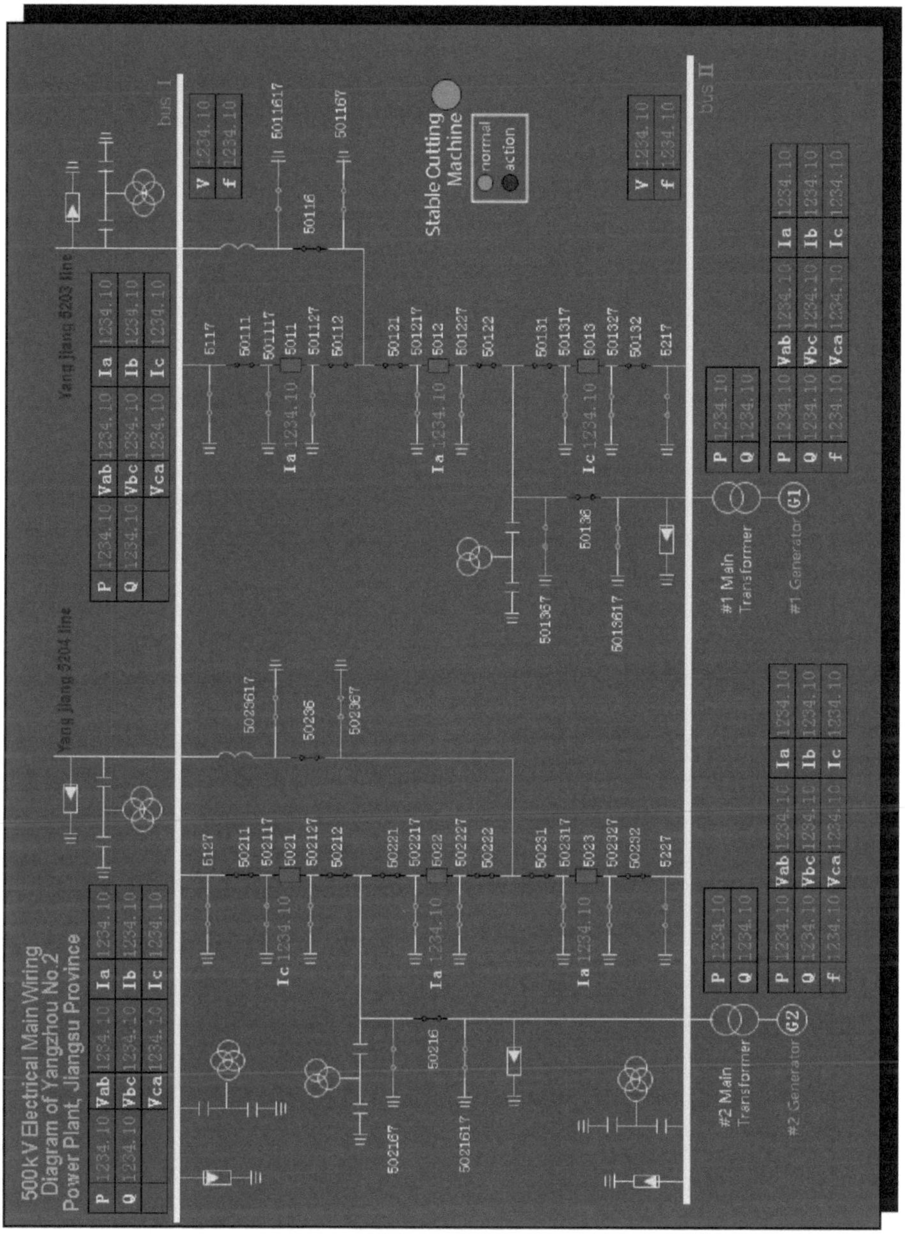

Figure 7.2: Example of a plant wiring diagram.

contour maps, and surface maps). These types of visualization methods are suitable for cases involving many buses and continuously distributed values. The methods also help the operator quickly grasp the distribution of the macroscopic data.

Line-type data refer to data associated with the transmission lines in power systems. Line-type data commonly include active power, reactive power, line loss, etc. The visualization of directional line-type data (e.g., active power flow) generally involves an arrow to indicate the direction, where the width of the arrow indicates the power flow value and the rectangular base of the arrow represents the thermal limit of a line's active power.

7.2.2 From state visualization to monitoring and control visualization

A smart power system is committed to realizing the multi-index optimality-approximating operation of power systems. To this end, a set of automatic operation systems for uninterruptedly monitoring the operation status of power systems is needed. The basic functions of the system include the following:

(1) Generate the corresponding dynamic indices based on the operating state of the power systems.
(2) Automatically generate events when the operating dynamic indices reach their limits.
(3) Automatically generate control commands to eliminate the generated events.
(4) Automatically generate operation instructions based on the control commands.
(5) Monitor the execution effect of the operating instructions.

Therefore, smart power system visualization should visualize not only the dynamic operating state but also monitoring and control, i.e, operating indices, events, control commands, operating instructions, and execution effects. Monitoring and control visualization helps operators intervene in automated systems and thereby combines machine intelligence and the artificial intelligence of system operators.

Monitoring and control visualization is a novel technology, and currently, little research has been carried out in this area. The following indices are discussed as brief examples.

First, different visualization patterns are designed for different indices of the dynamic operating state to indicate a more intuitive real-time operating state and to exert control over power systems. Table 7.1 lists common operating indices and their recommended form of visualization.

Second, the visualization of operating dynamic indices should not simply list the abovementioned visualized indices randomly but rather systemize the indices and organically integrate them into the corresponding topological graphics. Figure 7.3 illustrates the output of power generators and their remaining margins using three-dimensional cylinders on a single-line diagram.

Table 7.1: Common operating indices and recommended form of visualization.

Real-time operating dynamic indices		**Visualization**
Global indices	Frequency	Dial plate
	Grid loss	Two-dimensional curve
	Active power margin	Three-dimensional cylinder and pie
	Voltage margin	Stable domain curve
Zone indices	Zone active power margin	Three-dimensional cylinder and pie
	Zone reactive power margin	Three-dimensional cylinder and pie
	Tie-line active power	Three-dimensional arrows and pipelines
	Zone load forecasting	Two-dimensional curve
Element indices	Generation unit output	Three-dimensional cylinder and pie
	Bus voltage	Three-dimensional color rendering chart
	Thermal limit	Flash warning
	Remote signaling modification	Fill flicker
	Interface transmission capacity	Three-dimensional cylinder and pie

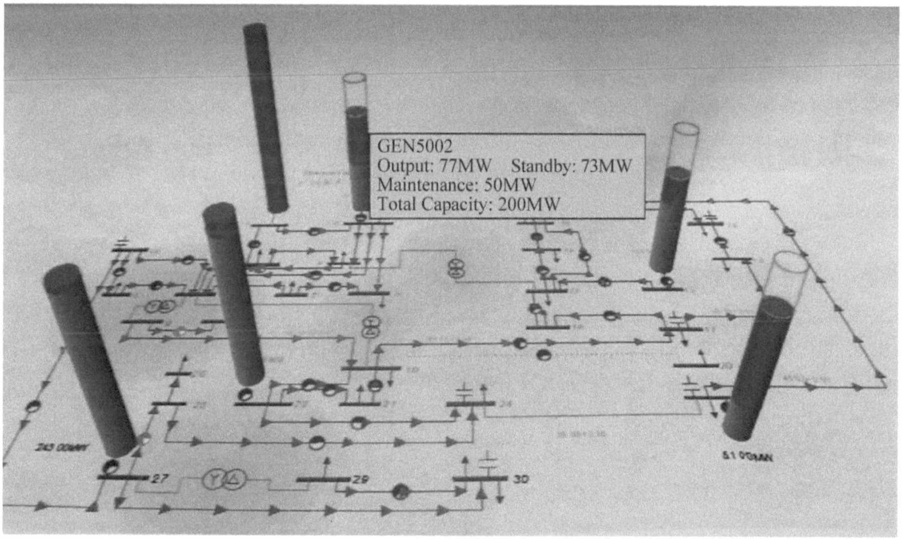

Figure 7.3: Example of operation index visualization.

7.3 Automatic generation of topological graphics

Graphical visualization of power systems refers to the visualization of power system topology, operating state, and other information by using graphics or images to help operators gain a timely and accurate understanding of the power system operating state [63–65]. Such visualization represents an important platform and means for operators to monitor, analyze, and regulate power systems. Traditional graphics are manually generated and maintained and have the following deficiencies:

(1) Traditional graphics require a great effort in development and maintenance and can lead to committing errors; moreover, the burden and cost of development and maintenance increase with the continuous expansion of the system scale.
(2) Interoperability is difficult. Due to the use of inconsistent graphic formats and different levels of information security, the interaction of graphic information between the different application systems is very difficult, and a set of graphic systems must be maintained. Maintaining such systems not only increases the cost of development and maintenance further but also causes data redundancy, information inconsistencies, and inconvenience to daily operation.
(3) Upgrading graphical display modes is difficult. If a change is made in a graphical display (e.g., a three-winding transformer drawing method is converted from one form to another), all graphics need to be modified.

To solve the issues of interoperability, the literature [66, 67] has proposed a graphical interactive standard form using Scalable Vector Graphics (SVG). SVG is an open standard text vector language for graphic description and can theoretically describe any complex graphics. However, there is no objective definition for electrical equipment; therefore, each manufacturer has a different element description of electrical equipment. This variation makes it difficult to convert graphic information between different application systems, even if the systems use the same SVG form.

The primary cause of the abovementioned issues is the relative isolation of the corresponding model and graphics. If a graph can be automatically generated based on a model, all the above-mentioned issues can be solved. Topological graphics are the basis and carrier of the visualization of power system information, which must be generated first. A brief description is provided below.

7.3.1 Basic concepts

The basic concept underlying the automatic generation of topological graphics is that a graph is automatically generated according to a model, specifically, the Common Information Model (CIM).

The CIM is the standard issued by the International Electrotechnical Commission (IEC); this standard defines all objects and their associated relationships in power companies. The CIM contains all the information indicated in the corresponding graphics (e.g., the topology and wires in the CIM), which contain the model data pertaining to the power devices and their connections in power systems.

All topological graphics (including single-line diagrams and plant wiring diagrams) are automatically generated based on the CIM. A GIS-based single-line diagram can also be generated if GIS information is available. Figure 7.4 presents the frame diagram of the automatic generation system of topological graphics. This figure shows that the data platform provides the necessary information for the automatic generation of topological graphics in the form of standardized interfaces, including network topology and real-time operation information.

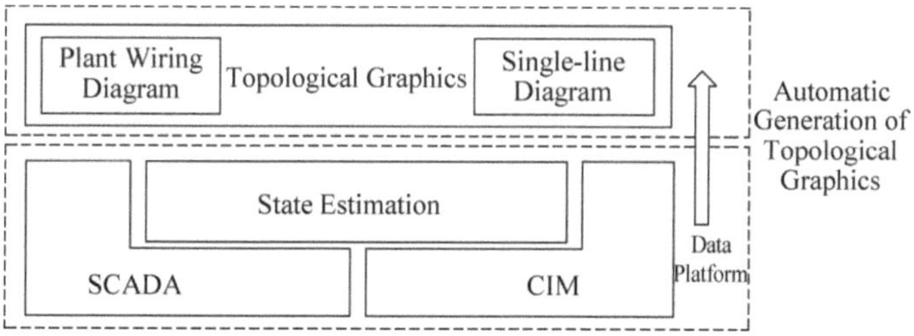

Figure 7.4: Frame diagram of system platform.

Automatic generation of topological graphics can not only guarantee the real-time tracking of the current topology and operating state of the power system, achieve self-maintenance of graphics, and save greatly on manual development and maintenance costs, but also allow the graphics and system models to be more closely coupled and interacted.

Automatically generated graphical information is actually a collection of graphic primitives and their coordinate information; therefore, the generated graphics can be displayed on any platform and do not depend on the graphical tools used by the platform. The display style and the visualization effect of generated graphics are unified. If one is changed, the other will also be changed. Therefore, the display is easy to customize according to the user's needs. Moreover, graphical information can also be exported to any graphic format. Given the abovementioned characteristics of graphics tools, a variety of advanced visualization technologies (e.g., three-dimensional visualization technology) can be easily imported.

The concept of automatic generation of topological graphics provides a new perspective for the visualization of power system graphics and provides advanced technical support for developing smart operations. The following sections introduce methods for automatically generating topological graphics for single-line diagrams and plant wiring diagrams.

7.3.2 Automatic generation of single-line diagrams

Automatic generation of graphics is widely used in the automatic circuit layout of large-scale integrated circuits. The main research directions include the automatic assignment of bus locations and the automatic routing of lines. Previous studies [68, 69] have used the penalty function method for automatically assigning printed circuit boards of ultra-large-scale integrated circuits. With respect to the automatic routing of lines, common methods include the Lee algorithm [70, 71] and the Hightower algorithm [72]. These algorithms can ensure that the lines neither intersect nor go through the bus rectangle. The drawback is that the algorithms often create serious issues of multiple line segments to avoid the intersection of lines and cannot even complete connection in some cases.

With respect to the automatic generation of single-line diagrams, R. Klump, D. Schooley, T. Overbye et al, proposed a method for automatically generating single-line diagrams based on geographic information [73]. Y. S, Ong, H. B. Gooi, and C. K. Chan proposed a method for generating a single-line diagram for a distribution network [74]. On this basis, B. Qiu and H. B. Gooi developed a network-based system for automatically generating single-line diagrams of distribution networks [75]. This method can yield improved results when applied to distribution networks with a radiation structure; however, it is not suitable for generating single-line diagrams of transmission networks with a complex network structure.

To overcome this shortcoming, we propose a new method for automatically generating single-line diagrams for transmission systems. The following sections introduce the bus layout, line routing algorithms, and their actual applications thereof. The software package developed in this book has been used in the actual operation of power systems as a part of SEMSs.

7.3.2.1 Bus layout

(1) Problem model

A layout P has a total connection length $L(P)$. After selecting the minimum total length of the connection as the optimization target, the objective function is as follows:

$$L(P) = \sum_{i=1}^{n} \sum_{j=1}^{n} W(i,j) d[P(i), P(j)] \tag{7.1}$$

where n is the number of buses in the network, $W(i, j)$ is the weighting factor of the connection between bus i and bus j, and $d[P(i), P(j)]$ is the distance between two buses.

In actual calculations, the number of lines between bus i and bus j is often used as the weighting factor, and the distance $d[P(i), P(j)]$ between bus i and bus j is calculated by equation (7.2):

$$d[P(i), P(j)] = |x_i - x_j| + |y_i - y_j| \tag{7.2}$$

where (x_i, y_i) and (x_j, y_j) are the coordinates of bus i and bus j, respectively, on a two-dimensional plane. Therefore, the objective function of the optimization can be obtained as follows:

$$\min \sum_{i=1}^{n} \sum_{j=1}^{n} W(i,j)\left(|x_i - x_j| + |y_i - y_j|\right) \tag{7.3}$$

(2) Solution algorithm

The abovementioned problem is a large-scale combinatorial optimization problem, which is difficult to solve. Therefore a heuristic method is adopted, which involves obtaining the initial layout of the buses and then further optimizing the initial layout.

(i) Initial layout

The entire layout space is divided into a series of grids. A grid corresponds to a position in the layout space in which only one bus can be arranged.

The distance $d[P(i), P(j)]$ between bus i and bus j is redefined as follows:

$$\begin{cases} |x_i - x_j| + |y_i - y_j| & , \text{ Buses } i, j \text{ are in the feasible region} \\ \infty & , \text{ others} \end{cases}$$

Furthermore, the entire layout space is used as the feasible region of layout P. Assuming that the initial positions of all buses are in the feasible region of layout P, the following algorithm can be used to obtain the initial layout of a single-line diagram. First, a bus is moved to the feasible region; then, for any bus that is still outside the feasible region, we search the entire feasible region, move it to an unoccupied grid in the feasible region, and make the penalty function reach the smallest value.

(ii) Layout optimization

The number of search levels is set to λ, and the target positions are selected as $A_1, A_2, \cdots, A_\lambda$. All contents of the target locations $A_1, A_2, \cdots, A_\lambda$ are exchanged. If the objective function value drops after the exchange, the exchange with the largest decreasing value of the objective function is set as the formal exchange to form a new layout. If the objective function value does not decrease during the exchange, the original layout is maintained. The abovementioned steps are repeated until all the buses are arranged.

In practical applications, it is appropriate to select two as the search level of the algorithm to result in a faster calculation speed. The greater the number of search layers that are chosen, the more time the algorithm requires and the better optimized the objective function will be.

7.3.2.2 Line routing

After all the bus locations are determined, it is necessary to determine the connections between the buses. Different types of buses, for example, power plants and substations, may have different shapes. However, because the bus shape has little effect on line routing, to simplify the problem, the specific shapes of the buses are ignored, and the buses are represented by rectangles in the study of line routing.

Unlike in large-scale integrated circuits, power systems allow for a few line intersections in single-line diagrams. For better visualisation effects, several principles should be followed: (1) the number of bent lines should be as small as possible, (2) lines should not cross the rectangles representing buses, (3) no two lines should overlap, and (4) the number of line intersections should be as small as possible. Based on these principles, the corresponding line routing methods are given as follows, including the determination of routing modes and the avoidance method when two lines overlap.

(1) Determination of routing modes

For clear and concise illustrations, only three connection methods can be used: direct routing, double polylines, and four polylines, as shown in Figures 7.5–7.7, respectively.

(i) Direct routing

Direct routing is applied when two rectangles are vertically or horizontally adjacent; a line can be used to connect the two rectangles.

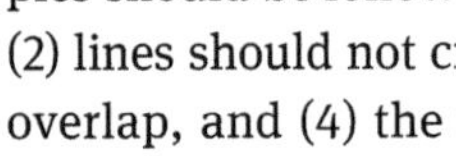

Figure 7.5: Direct routing.

(ii) Double polylines

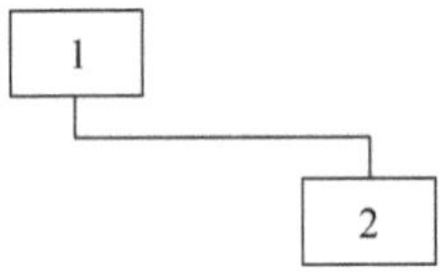

Figure 7.6: Double polylines.

Double polylines are applied when two rectangles are not vertically or horizontally adjacent; double polylines can achieve any type of connection between two rectangles.

(iii) Four polylines

Four polylines are used for bus avoidance. As shown in Figure 7.7, Bus 1 and Bus 2 can be directly linked. For avoiding intersections with Bus 3, four polylines are used.

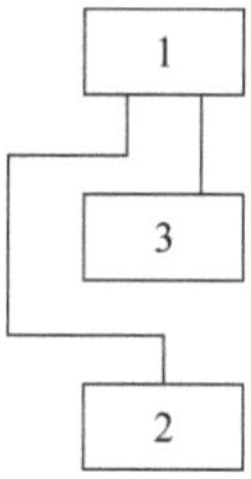

Figure 7.7: Four polylines.

(2) Routing avoidance algorithm

It is necessary to perform routing avoidance to prevent lines from overlapping, including vertical avoidance or horizontal avoidance. The specific algorithm is as follows:

(i) Use two avoidance arrays and save the horizontal and vertical paths in the graphics that have been arranged during line routing.

(ii) According to the method proposed in "(1) Determination of routing modes," the new line routing is arranged.

(iii) The avoidance arrays are checked, and it is determined whether the new lines and the original lines overlap. If there is any overlap, left and right or up and down, fine-tuning is applied to the positions of the new lines, and it is then determined whether the graphic still requires avoidance. If so, the adjustment is repeated until no avoidance is needed.

(iv) The routing path of the new line is determined and stored in the avoidance array.

In summary, the proposed line routing method is a heuristic method, and the calculation burden is much smaller than that of the Lee algorithm, the Hightower algorithm, etc. Although there may be few line intersections in the generated graphics, the intersections do not significantly affect the aesthetics and readability of the graphics.

7.3.2.3 Application example

Single-line diagrams of the IEEE 14-bus system and IEEE 39-bus system are arranged in this section. The diagrams of the IEEE 14-bus system and IEEE 39-bus system are shown in Figures 7.8 and 7.9, respectively.

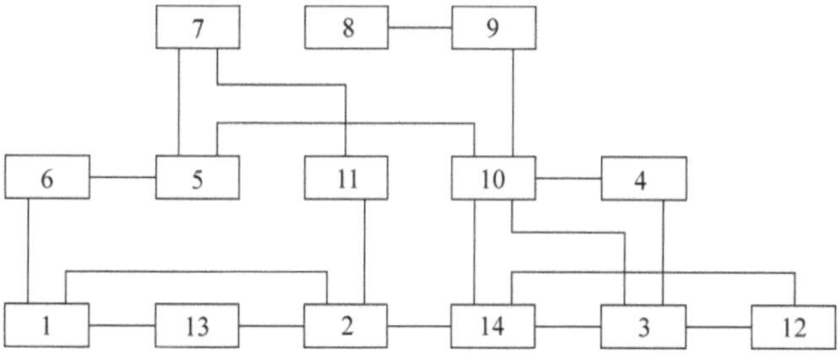

Figure 7.8: Automatically generated single-line diagram of the IEEE 14-bus system (19 lines).

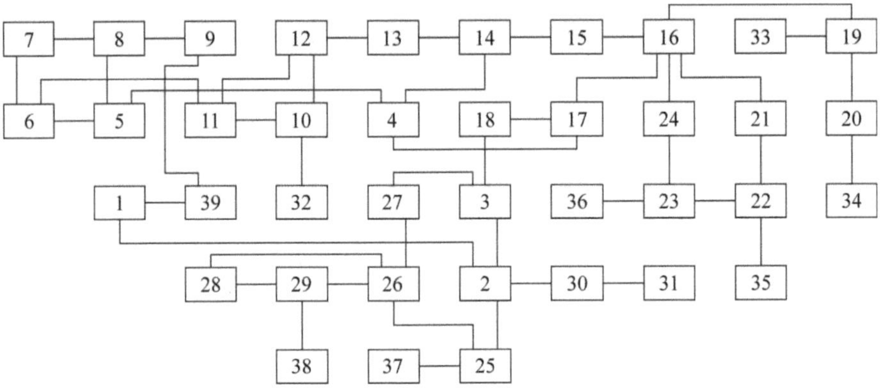

Figure 7.9: Automatically generated single-line diagram of the IEEE 39-bus system (46 lines).

7.3.3 Automatic generation of a main plant wiring diagram

A plant wiring diagram is the logical expression of the device connections in the main network. Based on system reliability and security, a main plant wiring diagram generally occurs in the form of typical wiring; that is, the equipment layout mode in the wiring diagram is fixed for the same type of wiring. Therefore, the automatically generated program design uses the method of "matching typical template." The program procedure is illustrated in Figure 7.10 and described as follows:

(1) Define the template for a typical wiring form

The template contains information on both internal attributes and external interfaces. Internal attributes are abstract descriptions of the arrangement rules of the wiring form, which represent the relative positions of devices (e.g., buses, circuit breakers, and transformers) and the corresponding forms of connection. The external interfaces are used to describe the individual characteristics of the corresponding wiring forms. Template matching conditions are included and exhibit mutual exclusion and exclusivity.

(2) Analyze the CIM

By analyzing the CIM, the relevant connections are obtained. The CIM is read using the Eclipse Modeling Framework. The device connection substring in a plant is obtained by using the depth-first search algorithm.

(3) Match the typical wiring form

First, the plant is divided into a number of topological islands based on the voltage level. Then, by plug matching using a socket, according to the connection characteristics of the devices in the topological islands, the coincident "template" is found to complete the match.

(4) Generate the position coordinates of the devices

According to the matching "template" and the layout rules of its internal properties, the equipment is arranged on a topological island, and the corresponding grid coordinates are generated. The display area is divided based on the number and structure of the topological islands, and the grid coordinates of the devices are mapped to the corresponding display area.

(5) Draw a plant wiring diagram

When a graphic primitive and its coordinate information are known, a plant wiring diagram is drawn using a graphical development language.

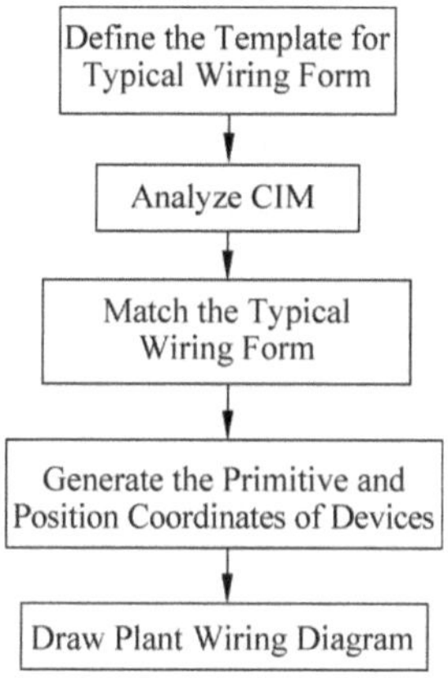

Figure 7.10: Algorithm for automatic generation of the plant wiring diagram.

The following two wiring forms (two-bus and four-segment, two-bus and four-segment with dual-bypass buses) are used as examples to illustrate template matching. The matching conditions and arrangement principles for the program design are listed in Table 7.2.

Table 7.2: Example of template matching.

Wiring form	Matching condition	Layout principles
Two-bus and four-segment	(1) Four buses (2) No bypass buses (3) Two buses have the same number of branches (4) Branches are directly connected to the corresponding two buses	(1) Bus segments are horizontally parallel. (2) Two buses are arranged parallel. (3) The layout of specific equipment is based on single-bus segmentation rules and double-bus rules. (4) The segment breaker is in the middle, and the breakers connected to buses are arranged up and down on the left and right sides.

Table 7.2 (continued)

Wiring form	Matching condition	Layout principles
Two-bus and four-segment with dual-bypass buses	(1) Six buses (2) Has two bypass buses (3) The remaining four buses should satisfy the matching condition of the wiring form "two-bus and four-segment."	(1) Two bypass buses are horizontally parallel on the top side. (2) The switch connected to the bypass bus is located below the bypass bus. (3) The rest buses and equipment should be arranged based on the rule of two-bus and four-segment.

According to the template defined in Table 7.2 and according to the algorithm illustrated in Figure 7.10, the wiring diagram is generated automatically. Figure 7.11 is an example of an automatically generated wiring diagram of the two-bus and four-segment form with dual-bypass buses.

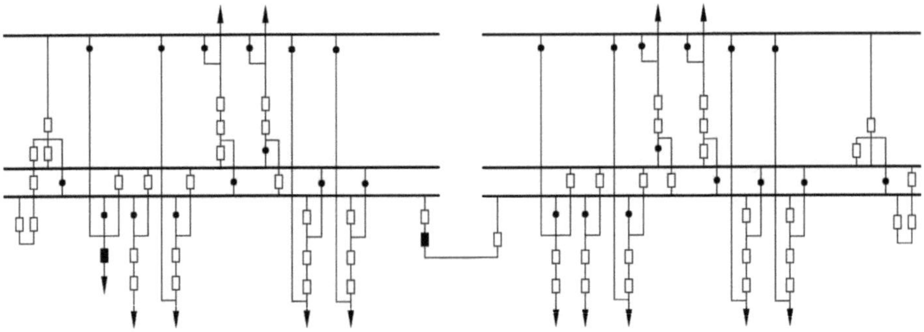

Figure 7.11: Example of an automatically generating plant wiring diagram.

Figure 7.12 shows the software structure of the automatic generation of the plant wiring diagram. The software is divided into four layers: the template layer is responsible for generating the template library containing the typical wiring forms; the data-reading layer is responsible for reading the CIM files and the real-time data; the analysis layer searches the topology information and matches the typical template, and the drawing layer generates the graphic primitive and its coordinate information, and draws the graphics.

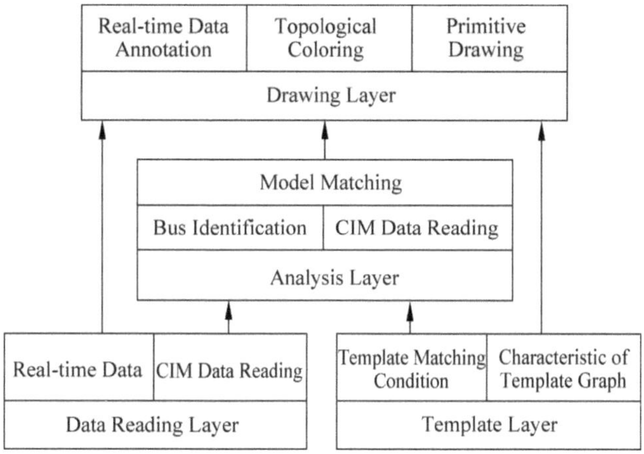

Figure 7.12: Software structure of the automatic generation of a plant wiring diagram.

7.4 Fast graphic drawing algorithm

This section describes a fast graphical rendering method, namely, an algorithm for the voltage contouring interpolation and the graphic generation based on grid merging. Although this idea is proposed for voltage contouring, it can also be applied to other similar situations.

Currently, power system operators must often handle a tremendous amount of bus data and cannot meet requirements based only on the visualization of form and text [76, 77]. With the development of computer visualization technology in recent years, voltage contouring based on geographical information has gradually received increasing attention from electricians. Voltage contouring generally uses color mapping to show the voltage value and treats the voltage magnitudes of key buses as a continuous distribution in the entire system coverage area. Psychology experiments [78] have shown that the color of voltage contours can arouse more attention from the human eye than a traditional digital display and is more intuitive in expressing the overall voltage level of each grid area. Thus, voltage contours can help operators find overvoltage buses and areas in a timely manner.

The principle of voltage contouring is to extend the geographically distributed discrete data to the entire drawing plane by certain interpolation algorithms. The current common interpolation algorithms for voltage contouring can be divided into two types: methods based on irregular triangulation and methods based on regular grids. Surface generation based on irregular triangulation is carried out in several steps: Delaunay triangulation, triangular edge equivalence tracking, curve fitting, patch filling, etc. In the triangulation method, although the triangle is optimized, it is inevitable that there are triangles with sharp edges and corners; thus, the sharp corners of curves may appear in the contours. The contours do not make smooth transitions easily, and the graphics are often less visually pleasing than those formed by the method based on a regular grid [80]. The method based on a regular grid divides the drawing screen into $M \times N$ uniform grids. The size of each grid depends on the value of the neighboring data points, with denser data points exhibiting greater weights. This method yields more visually pleasing graphics.

Currently, the main issue affecting the actual application of voltage contours in power system operation is the processing speed. Both interpolation and graphic generation processes are time-consuming. For example, based on the current common regular grid methods for completing a single diagram with a resolution ratio of 800 × 800, using a workstation with a 2.0 GHz frequency CPU and a 2 GB memory cost of 1–3 s, a single interpolation generally requires 0.5–1 s to complete, and a single diagram generally takes 0.5–2 s to generate. This speed is far from the visualization requirements of the power system dynamic process and may adversely affect the operating efficiency of other programs on the operation stations. Therefore, improving the speed of voltage contouring interpolation and graphic generation is an issue whose solution may provide high engineering value. A novel approach is proposed by reducing the number of meshes that need to be interpolated and drawn. This method can significantly improve the display speed of voltage contouring.

7.4.1 Interpolation algorithm analysis

Assuming that a display area is evenly divided into $M \times M$ fixed-size grids, for any grid point p on the plane, the task of the spatial interpolation algorithm is to calculate the virtual value of p based on the distance from p to the actual electrical bus and the voltage value of the actual electrical bus. The most classic interpolation algorithm is the method involving the inverse of the square of the distance [81], which is calculated as follows:

$$v_p = \frac{\sum_{i=1}^{N} \frac{v_i}{d^2{}_{\text{pi}}}}{\sum_{i=1}^{N} \frac{1}{d^2{}_{\text{pi}}}} \tag{7.4}$$

where N is the actual number of buses, v_i is the voltage magnitude of bus I, and d_{pi} is the distance from i to p. The reciprocal of the square is used as the weight because the square calculation is faster than the calculation of other distance norms. To reduce the computational burden, one study [81] proposed considering only the influence of the data points located at a distance less than d_{inf} (d_{inf} is the preset distance of influence) from p. Moreover, using the symmetry of the integer coordinate square grid, the distance calculation is reduced to 1/8 of the original calculation.

To overcome the issues of noncontinuous computational results and poor visual effects along the boundary of the affected area associated with the abovementioned algorithm, one study [82] proposed a novel interpolation formula:

$$v_p = \frac{Cv_\infty + \sum\limits_{\substack{i=1 \\ d_{pi} < d_{\text{inf}}}}^{N} \left[\frac{1}{d_{\text{pi}}^2} - \frac{1}{d_{\text{inf}}^2}\right] v_{\text{i}}}{C + \sum\limits_{\substack{i=1 \\ d_{pi} < d_{\text{inf}}}}^{N} \left[\frac{1}{d_{\text{pi}}^2} - \frac{1}{d_{\text{inf}}^2}\right]} \tag{7.5}$$

where $v_\infty = \frac{\sum_{i=1}^{N} v_i}{N}$ and C is a smoothing constant, which can adjust the smoothness of the interpolation results along the boundary of the affected area.

The two interpolation algorithms mentioned above can be expressed as follows:

$$v_p = \sum_{i=1}^{N} \alpha_{pi} v_i = \boldsymbol{A}\boldsymbol{V_N} \tag{7.6}$$

where α_{pi} is the influence weight of data point i to p. The value is a constant, related only to the relative distance, and is independent of the voltage; thus, the value must be calculated only once upon initialization. $\boldsymbol{V_N}$ is the vector of bus voltages; that is, $\boldsymbol{V_N} = [v_1 \quad v_2 \quad \ldots \quad v_N]^T$. A is an $M^2 \times N$ constant matrix, and α_i and α_j are the ith and jth row vectors, respectively, of the A matrix (called the weighting vector in this book). Because the affected area of the grid to be calculated is limited and the weighting vectors are generally sparse, sparse techniques can be used for storage and computation.

The voltage difference $\boldsymbol{dV}_{ij}$ between two adjacent grids i and j is

$$\boldsymbol{dV}_{ij} = (\alpha_i - \alpha_j)\boldsymbol{V}_N = ||\alpha_i - \alpha_j|| \cdot ||\boldsymbol{V_N}|| \cdot \cos\theta \tag{7.7}$$

where $||\alpha_i - \alpha_j||$ is the Euclidean distance between the two weight vectors and θ is the angle between the two vectors. The per-unit value of the bus voltage is generally believed to not exceed 1.1. Let $c = 1.1$ and $||V_N|| \le c = 1.1$. Then,

$$\boldsymbol{dV}_{ij} \le c||\alpha_i - \alpha_j|| \le \delta \tag{7.8}$$

where δ is the threshold. Thus, we conclude that for a sufficiently small value $||\alpha_i - \alpha_j||$, the difference between the voltage values of the two adjacent grids is always small, regardless of how the bus voltage vector V_N changes.

7.4.2 Grid merging method

The interpolation calculation and the graphic generation associated with voltage contouring are proportional to the number of grids M^2. Increasing the number of meshes can produce more delicate and visually pleasing diagrams but also steeply increases the calculation time. While this approach can significantly increase the calculation speed, the image quality is reduced, and jagged graphs are generated.

Another method for reducing the number of graphics to be calculated while achieving satisfactory image quality is to draw and calculate only the grids near the actual electrical buses and ignore others. The drawback of this approach is that it is difficult to determine the drawing boundaries, especially for dispersed buses or those clearly divided into several groups.

By observing the completed rendering graphics, two extreme cases can be found: (1) The distance between bus A and its multiple electrical buses is small. The numerical change in the grids between A and neighboring bus C is significant, as long as bus A slightly narrows the distance from bus C. (2) Other grids that are farther away from all the buses or near a certain bus exhibit a very small change in values or almost no change. Therefore, to generate graphics, different resolutions can be used to address different positions. According to equation (7.8), the voltage value after interpolation is necessarily less than δ, as long as the distance of the weight vectors of the adjacent grids times c is less than the threshold δ. Thus, grids with a small distance of adjacent weight vectors can be combined into a large grid, herein referred to as a cell group. Moreover, the entire area is covered with the color of the cell group to reduce the number of grids.

Because the grid merging result depends only on the distribution of the data point positions, it is necessary to perform only one merging during graphic initialization.

For better visual quality and efficiency, the grid merging algorithm should comply with the following principles:

(1) The merged cell group remains rectangular to facilitate drawing.
(2) The numerical difference between any two points in the merged cell group is smaller than the threshold δ.
(3) The total number of grids after merging should be as small as possible.

One efficient and practical grid merging algorithm is one in which only two traversals are required for all the original grids. The algorithm is as follows:

(1) Calculate the weight vector α_i of each cell of the original grid, based on the applied interpolation algorithm.
(2) For the ith cell C_i, calculate its weight vector distance among the four adjacent cells. If the weight vector distance between the cell and an adjacent cell times c is less than δ and neither cell belongs to any cell group, a new cell group G_k is created, and the two cells belong to the new cell group. The weight vector of the ith cell is selected as the weight vector of the new cell group.
(3) For the new cell group G_k, try to extend the group toward its surroundings. The priority extends outward from the longer side of the rectangle so that the expanded cell group can be as large as possible. Each cell at the new extended edge must not belong to any existing cell groups, and the weight vector distance of each cell and cell group at the new extended edge times c should also be less than δ.
(4) Repeat step 3 until no new edges can be expanded.
(5) Move to the I + 1 cell and repeat step 2.
(6) Finally, create a separate cell group for each unmerged cell.

Subsequent interpolation and graphic generation procedures must handle all the cell groups. The interpolation is performed based on the center of the merged cell group (the intersection of the diagonals); all the cell groups are colored after the diagram is generated. Because the number of cell groups is less than the initial number of cells, the computational burden will be reduced.

7.4.3 Application examples

Here, a six-bus system is applied as an example (shown in Figure 7.13) to illustrate the grid merging effect. Six buses are evenly distributed on a 2D plane.

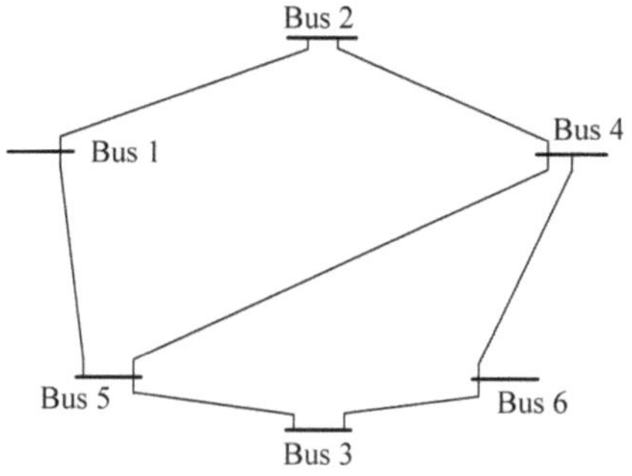

Figure 7.13: Single-line diagram of a six-bus system.

The weighting vector is calculated using the classical algorithm described by equation (7.5). Color mapping is based on the scheme described in [82]. To facilitate the display of the combined effect, the size of the initial rectangular grid is 80×80. In grid merging, the threshold is selected as $\delta = 0.0003$, based on experience, thereby ensuring that there is no visible trace of the change between adjacent patches. The result of grid merging is shown in Figure 7.14, where the blank area represents the unmerged cells; the rectangle of each black border represents the merged cell group. The figure shows that the cell groups close to a bus or away from all buses have the largest areas, and cell merging occurs less often in the transition zone between the buses. Compared with the number of merged rectangles in traditional meshing, the number of merged rectangles is significantly reduced.

Three initial grids measuring 80×80, 400×400, and 800×800 are selected to test the improvement in the image calculation efficiency of the novel algorithm. Table 7.3 provides a time comparison of the interpolation algorithm and the graphic generation. The table shows that the novel algorithm improves the speed dramatically.

Test environment: ThinkPad T60 notebook. CPU with a frequency of 1.83 GHz and 2 GB of memory. All procedures are written in Java. The graphical display uses Java OpenGL (Open Graphics Library) toolkit (JOGL).

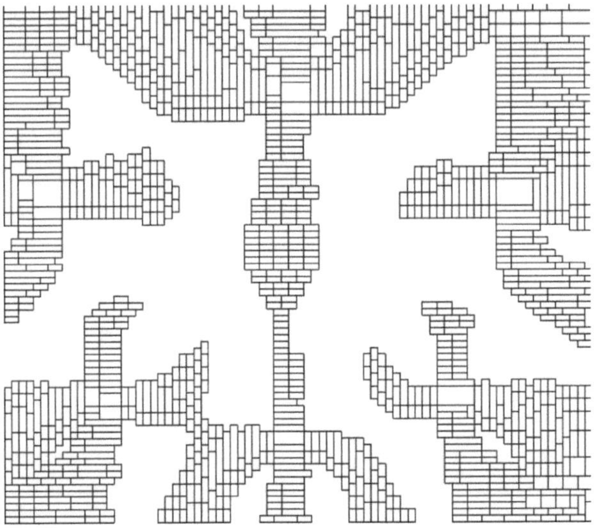

Figure 7.14: Merged cell group of a six-bus system based on 80 × 80 initial grid.

Table 7.3: Time comparison of voltage contouring algorithms on a six-bus system for different selected grid sizes.

Algorithm	Time required for interpolation calculation/ms			Time required for graphic generation/ms		
	80 × 80	400 × 400	800 × 800	80 × 80	400 × 400	800 × 800
Original algorithm	16	172	625	31	188	563
Improved algorithm	0	46	93	16	31	62

Table 7.4 compares the numbers of rectangles that must be processed before and after the algorithm is improved for different grid sizes.

Table 7.4: Comparison of the numbers of rectangles that must be processed in different algorithms (six-bus system).

Algorithms	Number of rectangles to be processed in interpolation and graphics generation		
	80 × 80	400 × 400	800 × 800
Original algorithm	6,400	160,000	640,000
Improved algorithm	4,136	23,417	49,995

Table 7.4 indicates that the advantage of the improved algorithm is more distinct for larger grids. As the grid size increases, the difference in the number of rectangles to be processed before and after the algorithm improvement becomes even greater. The difference reaches even one order of magnitude.

The grid merging algorithm improves the calculation speed and ensures the quality of the graphic display. As shown in Figures 7.15 and 7.16, the effect of grid merging does not decrease for the 400 × 400 grid.

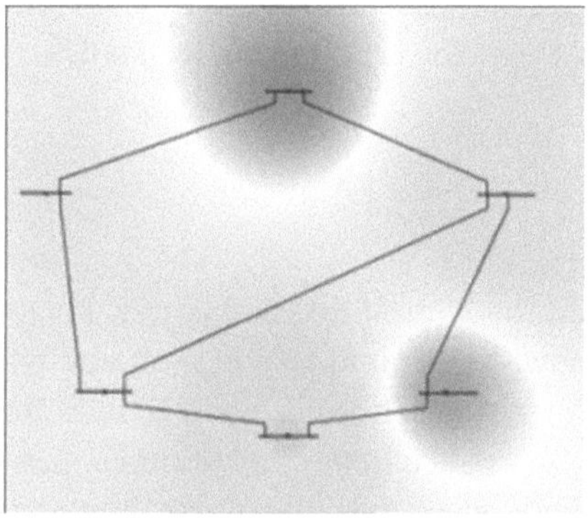

Figure 7.15: Display effect of the original grid.

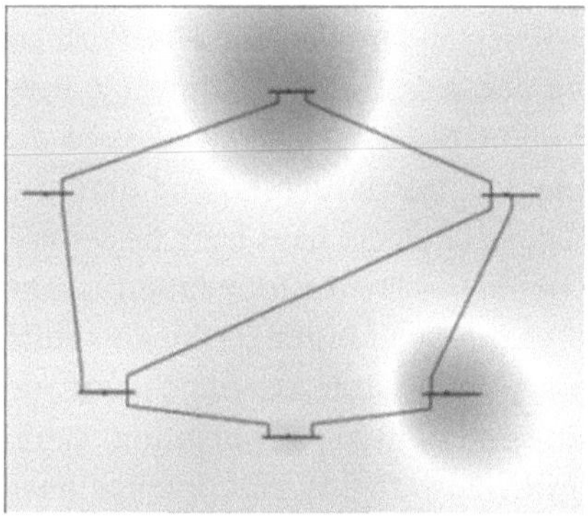

Figure 7.16: Display effect after grid merging.

Using actual data as an example, the geographical wiring diagram of the 500 kV and 220 kV voltage levels of a power grid with 113 plants and substations is examined. The test is conducted using an 800 × 800 rectangular grid. The results are shown in Table 7.5.

Table 7.5: Comparison of graphic calculation times on a 113-bus system of a power grid.

Algorithm	Time required for interpolation calculation/ms	Time required for graphic generation/ms	Number of rectangles to be processed
Original algorithm	1,031	573	640,000
Improved algorithm	375	188	154,118

Test environment: CPU with a frequency of 1.83 GHz and 2 GB of memory. All procedures are written in Java. The graphical display uses the Java OpenGL toolkit JOGL.

7.5 Summary

The operators of smart power systems still hold the highest decision power. Determining how to improve operators' "right to know" and "rapid response capability" are important issues in research on smart power systems. It is necessary to indicate the real-time operation of power systems to operators in a timely and accurate manner. Thus, smart power systems must develop advanced visualization technology. This chapter begins with an analysis of the visualization content in smart power systems and introduces the method of automatically generating topological graphics of power systems and plant wiring diagrams based on standard data models. Furthermore, to improve the drawing speed of 3D graphics, a fast graphics-drawing algorithm based on power system characteristics is proposed.

Thus far, the basic principles, key technologies, infrastructure, and platform design of smart power systems based on hybrid control theory have been presented. These novel ideas, theories, methods, and technologies lay the foundation for the construction of smart power systems. The next chapter will introduce how to build a novel operation automation system based on power hybrid control theory and other key technologies to realize the multi-index optimality-approximating operation of power systems and distribution networks. Both theory and practice have shown that smart power systems are practical and effective and will become the future direction of power system development.

Chapter 8
SEMS

8.1 Introduction

In the preceding chapters, the definition, theoretical foundation, infrastructure, basic support platform, and several key technologies of smart power systems (SPSs) were introduced. On this basis, this chapter discusses how to utilize the existing infrastructure and platform with the application of power system hybrid control theory and other key technologies to develop a novel supervisory control and data acquisition (SCADA) system, thus realizing **multi-index optimality-approximating operation** in power systems. We call this system the smart energy management system (SEMS). The introduction is presented as follows.

8.2 Definition and characteristics of SEMS

8.2.1 Definition of SEMS

There is no doubt that the modern power system is the largest, most complex electromagnetic system with maximum coverage on the planet. Currently, the control and dispatching of each power system rely mainly on the operator's knowledge and experience and are complemented by an appropriate energy management system (EMS) [83]. Most of the EMSs that are in operation in China were developed in the 1990s. Due to the limitations of computer and control technology at that time, the main function of these EMSs is to help dispatchers understand the conditions of the power system. In addition, some of the EMSs can make some suggestions for dispatch in some respects. However, real closed-loop automatic control has not been achieved. **Multi-index optimality-approximating control has** always been the long-term goal in theory and engineering.

Power system hybrid control theory provides a new methodology for **multi-index optimality-approximating control**. As described in Chapter 2, conditions that do not meet industry standards but satisfy the dispatcher are defined as events; thus, the generated events can be systematically analyzed and regulated. When all events are eliminated, the system returns to a satisfactory condition.

Based on power system **hybrid control theory,** united automatic cooperative scientific control can be achieved by SEMSs, within power systems, to control various controllable resources (including generators, capacitors, reactors, transformers, and a variety of compensation, flexible AC transmission system (FACTS), and large storage systems, etc. Alternatively, SEMSs can control various controllable resources (including

https://doi.org/10.1515/9783110448825-008

generators, capacitors, reactors, transformers and a variety of compensation, FACTS, large storage systems, etc.) within power systems unitedly, automatically, cooperatively, and scientifically) such that the power system is a multi-index optimal control automation system.

8.2.2 Characteristics of the SEMS

The SEMS is based on power system **hybrid control theory**, and the essence of this theory is event-driven. Therefore, SEMS has event-driven characteristics and attributes. Consequently, SEMS has the following characteristics:

The first characteristic is a digitized representation, which is the foundation of the entire system operation. A digitized representation includes two aspects: the first is a digital representation of a power system model, including physical conditions (i.e., static features) and operation states (i.e., dynamic features) of the power system; the second is a digitized representation of the control flow based on events, including the digital representation of events, control commands, operation commands, and the control process.

The second characteristic is global data sharing, which is the basis for realizing intelligent decision-making. For global sharing of data, it is necessary to build a SEMS for a power system and its interconnection system, with a data-sharing platform connected to its superior control center. Additionally, it is also necessary to establish appropriate mechanisms for data sharing within the system (i.e., using standard information models and a data access interface in accordance with the IEC6197 standard).

The third characteristic is intelligent global decision-making. On the one hand, "global" here refers to decision-making that is developed based on a global digital model. On the other hand, such decision-making is required to utilize overall control methods. Intelligent decision-making can largely replace the work executed by the dispatcher. The implementation of intelligent decision-making depends on the application of a new control theory and new analysis methods and algorithms in the power system.

The fourth characteristic is the regulatory mechanisms of the power system **hybrid control theory.** Event generation is the trigger for control actions, and event removal (removing events) is the target of control. This aspect is a simple mechanism to meet the demand for comprehensive optimization of complex systems. This mechanism not only simplifies the complexity of the problem but also ensures the optimal results of comprehensive optimization, which is difficult to achieve by other methods.

The fifth characteristic is the capability of powerful visualization. The SEMS graphically shows the real-time conditions of the power system to dispatchers. As a result, dispatchers can understand the situation of the system operation clearly, comprehensively, and promptly. Additionally, dispatchers may make decisions about whether, when, and how the system operation should be supervised. Thus, a high degree of integration of machine intelligence and human intelligence can be achieved, and the power system operation is improved.

In addition to the characteristics mentioned above, the SEMS also has some specific requirements for composing and implementing software systems.

The first requirement is a modular structure. For conventional changes (such as the type of event increases and changes in the event criterion) in the system, the composition framework of the system must remain relatively stable; thus, the system requires the flexibility to add a functional unit or modify and enhance an existing functional unit, without changing the overall framework. A modular structure, based on component technology, can not only reduce the time and costs of construction and maintenance but also provide convenient conditions for interoperability and interchange among functional modules within the system.

Second, the implementation of standardized models is applied for implementing the SEMS, mainly because the event analysis and event processing in the SEMS are based on digital models of electric power systems. These models describe part of or one aspect of the function, structure, and behavior of the power system in a normalized manner. Relatively, digital models of the power system do not change frequently, but the technologies applied to realize corresponding applications are being constantly improved. Thus, the use and design of "standardized models" are beneficial for tracking and adopting new technologies and enhancing the level of automation dispatching in the power system over time.

8.2.3 SEMS and EMS

The SEMS is not only compatible with a traditional EMS (TEMS) but will also coexist with a TEMS for a long time (approximately ten years). Many of the functions in SEMS can be achieved by inheriting and extending the corresponding functions in a TEMS. In fact, the relationship between both groups of functions can be expressed in one sentence: If and only if the power grid is in an event-free state, as judged by the SEMS, the power system operation and monitoring are executed by the TEMS.

8.3 Components of SEMS

As mentioned earlier, the essence of power system **hybrid control theory** is event-driven. Accordingly, the SEMS is composed of an event analysis system, event processing system, and decision-making system for dispatchers. The overall structure of the SEMS is illustrated in Figure 2.5. This section adds Figures 8.1 and 8.2 and appropriate instructions to analyze the subsystems, including information interaction, event analysis, and event processing, in detail.

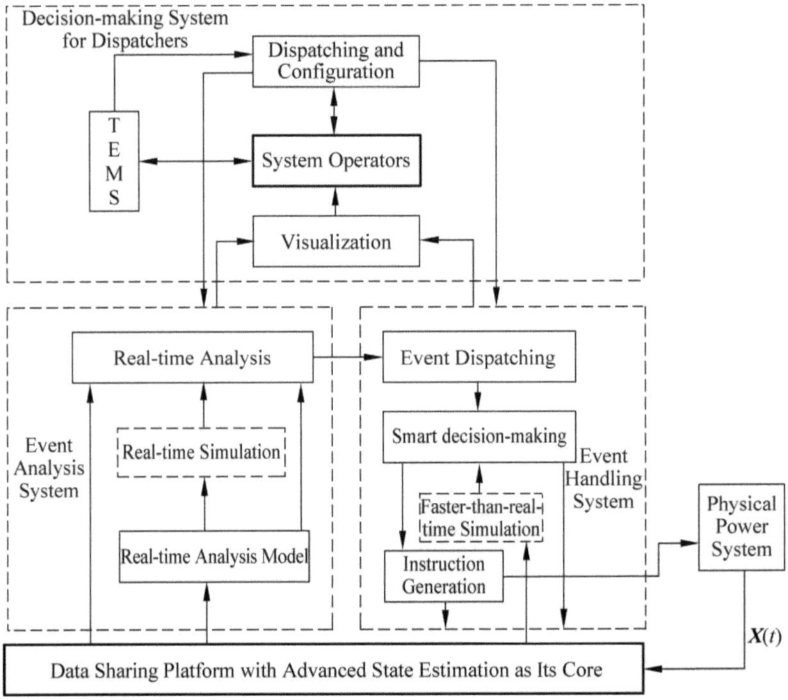

Figure 8.1: Diagram of the components and interrelationships in the SEMS.

(1) The event analysis system completes the real-time analysis of information obtained by the data-sharing platform to determine whether an event has occurred. If so, the event is sent to the event processing subsystem; otherwise, a TEMS is in charge.
(2) The event processing subsystem handles the received events and generates corresponding control commands, which will be assigned to the various types of regulation units.
(3) The data exchange among the abovementioned event analysis subsystem, event processing subsystem, and other subsystems is implemented by a "data sharing platform with advanced state estimation at its core." A data-sharing platform is the basic support platform of an SPS, as discussed in detail in Chapters 4 and 6.

Under normal conditions, effective closed-loop control can be carried out by developing the above subsystems and organizing them according to Figure 2.3. Because it is essential for dispatchers to have a good "degree of knowledge" about the regulated power system operating status and to be able to perform timely interventions when needed, a decision-making system for operators is also included in the SEMS. The main function of this decision-making system is to visualize the status of the system operation for dispatchers so that they can intuitively, clearly, and comprehensively understand the system's operating conditions, and conveniently implement dispatching and configuration operations.

8.3.1 Event analysis system

The task of the event analysis system is to complete real-time analysis for the true state obtained by the data-sharing platform, thus judging whether an event occurs and what kind of an event it is. The methods are as follows:

(1) Based on the analysis, determine whether the index of security and stability (Level I), the index of power quality (Level II), and the index of low energy consumption (Level III) are satisfied (described in Section 5.2.2).

(2) If index z is not satisfied, event E_z is generated as described in Section 2.4; if all indices are satisfied, the operating power system is **in a multi-index optimal state space** without any event, according to Definition 2.6.

The event analysis system consists of a real-time analysis model, real-time simulation, real-time analysis and judgment, and the input of a standard metrics system, as illustrated in Figure 8.2.

8.3.1.1 Real-time analysis and judgment

This aspect consists of a group of modules that can analyze and evaluate (by performing, e.g., network loss analysis, operation quality analysis, perturbation analysis, and security and stability analysis). These modules obtain real-time physical power system operating states from the real-time analysis model or real-time simulation system. By analyzing these data and comparing the results with the standard indices, it can be determined whether all the indices of security and stability (I_I), power quality (I_{II}), and low energy consumption (I_{III}) are satisfied, thus determining whether an event e_{zi} ($i = 1, 2, \ldots, r$) (see equation (2.10)) occurs and is submitted to the event processing system.

8.3.1.2 Real-time analysis model

For different types of events, it is necessary to establish different real-time analysis models. These analysis models contain a variety of data required for real-time simulation and analysis, and that is described by a standardized modeling language. The data involved include (1) network topology model data, real-time measurement data, and high-reliability state estimation data; (2) load distribution model data, which are provided by high-precision ultra-short-term load forecasting and bus load forecasting modules; and (3) a unit output plan (including ancillary services), which is generated based on the operating plan combined with trading data of the real-time market.

8.3.1.3 Real-time simulation

The real-time simulation is indicated by the dashed lines in Figure 8.2 and is a required functional module to implement advanced stages of the SEMS. The functions of the simulation include the following three aspects: (1) real-time simulation based on online data, the results of which are used for the real-time analysis system; (2) showing the results of the simulation to dispatchers using a visual interface, thereby providing references for dispatchers to generate dispatching instructions; and (3) serving as a unified computing platform for real-time analysis, and providing efficient and reliable computing analysis capabilities for different online analysis modules.

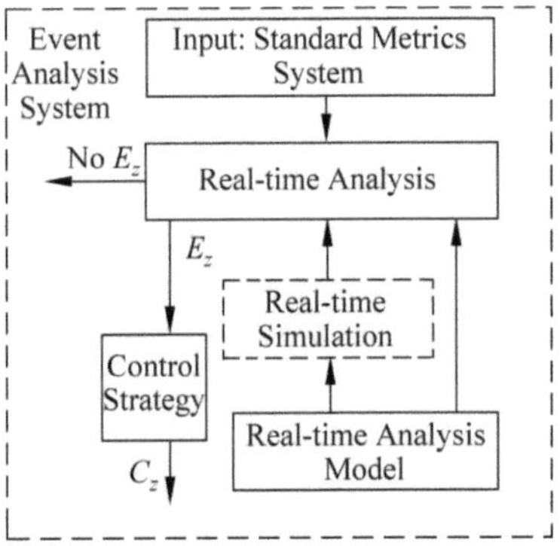

Figure 8.2: The components of the event analysis system.

8.3.2 Event processing system

The tasks of the event processing system are as follows: receive the set of events generated by the event analysis system $\mathbf{E}_i(i=1,2,\cdots,r)$; complete an ε transformation, based on the obtained true state set $\mathbf{X}_{TS}(t)$ and a reasonable, effective algorithm, $C_i=\varepsilon(\mathbf{E}_i)$; and transform $\mathbf{E_i}$ into a corresponding control command C_i. After the formation of C_i, the event processing system will proceed with a second-stage transformation – C transformation, $\mathbf{O}_i=C(C_i)$, thus generating a group of operations $\mathbf{O}_i=\{\mathbf{O}_{i1},\mathbf{O}_{i2},\cdots,\mathbf{O}_{iw}\}$, so that event $\mathbf{E}_i$ can be eliminated ($\mathbf{E}$ becomes an empty set) within a sufficiently short time Δt, and the physical power system is restored to a no-event state. Certainly, in some conditions where event processing is not very complicated, it is also acceptable to combine the ε and C transformations into one transformation, the composite expression of which is $\mathbf{O}_i=C(\varepsilon(\mathbf{E}_i))$.

The event processing system consists of modules, including event dispatching, smart decision-making, command generation, and faster-than-real-time simulation, as illustrated in Figure 8.3 and introduced as follows.

8.3.2.1 Event dispatching

The event-dispatching module receives events from the event analysis system. These events are processed by sorting (different events have different priorities; e.g., safety events must be processed before economic events) and merging (since handling a certain type of event may eliminate another type of event), and will be submitted to the smart decision-making module for further analysis.

8.3.2.2 Smart decision-making

The smart decision-making subsystem is in charge of the E transformation, that is, generating C_z for the corresponding E_z. This subsystem consists of a group of submodules that can generate corresponding control commands C_z to a mid-level subsystem (including the command generation module within the SEMS and controlled devices located in power plants and substations) based on different types of events E_z.

8.3.2.3 Command generation

Command generation is responsible for the C transformation, that is, producing a group of operations $\mathbf{O}_{zi} = \{\mathbf{O}_{zi1}, \mathbf{O}_{zi2}, \cdots, \mathbf{O}_{ziw}\}$ (see equation (2.12)) so that event E_z can be eliminated (E becomes an empty set) within a sufficiently short time Δt. Command generation includes a group of submodules that can generate a group of operating instructions $\mathbf{O}_{zi} = \{\mathbf{O}_{zi1}, \mathbf{O}_{zi2}, \cdots, \mathbf{O}_{ziw}\}$ for devices for different kinds of control commands C_z; thus, the operating instructions can be assigned to the appropriate receiving and execution devices. There are two forms of C transformation, as described in Section 2.5: the first is "centralized," meaning that the transformation is completed within the EMS centralized control center; the second is "distributed," meaning that the transformation is completed in power plants and substations.

8.3.2.4 Control test

The control test platform (CTP) (see Section 1.2.3), which is supported by the technology of faster-than-real-time simulation, is responsible for testing the performance of all operating instructions. By performing a faster-than-real-time simulation with control commands, C_z generated by smart decision making, the effect of control can be predicted; the effective commands are executed immediately, and the ineffective commands are amended, retested, and executed. This process ensures that each group of operations $\mathbf{O}_{zi}$, in accordance with the control command C, is valid. Certainly, the final output of the platform should be presented to dispatchers as clearly as possible.

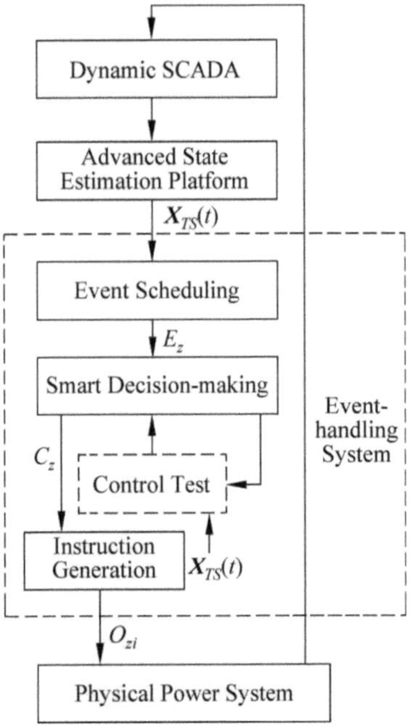

Figure 8.3: The structure of the event processing system.

8.3.3 Decision-making system for dispatchers

Dispatchers still have supreme decision-making authority, even after the realization of the SEMS, which is a highly intelligent and automated system. Dispatchers have two main duties: making dispatching decisions in the operating environment and configuration management in the maintenance environment. For example, although no event occurs when the system is operating, the operators can understand some potential problems (these potential problems may not cause an event) using the visual system and implementing dispatching operations proactively (the effects of dispatching operations can be predicted using the faster-than-real-time simulation system at this point; thus, the risky operation can be reduced).

In the SEMS, dispatchers' supreme decision-making authority is illustrated by the decision-making system for dispatchers; this system consists of two modules: visualization, and dispatching and configuration, as shown in Figure 8.4.

8.3.3.1 Visualization

The visualization module is responsible for the visualization of power system operation monitoring; that is, in addition to the dynamic visualization of operating conditions, the performance indicators, events, control commands, instructions, and implementation of the results must also be visualized. The monitoring visualization helps the dispatchers understand the system operating conditions comprehensively

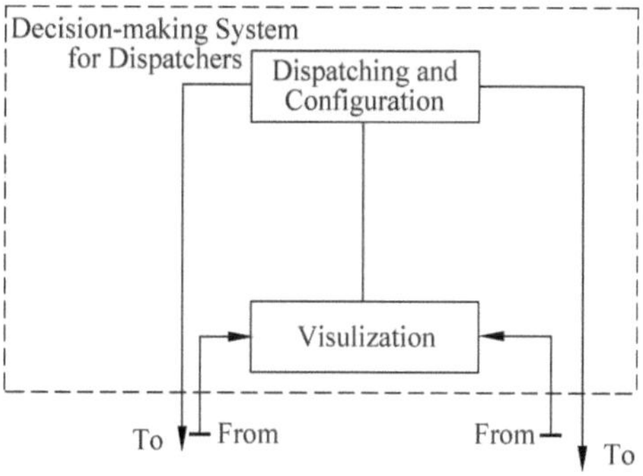

Figure 8.4: Diagram of the decision-making system for dispatchers.

and directly; thus, they can regulate the automatic dispatching system operation precisely and flexibly so that machine intelligence and the operator's intelligence can be united (integrated) very well.

8.3.3.2 Dispatching and configuration

The functions of the dispatching and configuration module can be divided into two parts: dispatching and configuration. Dispatching includes active and passive forms. For active dispatching, the dispatcher makes predictions based on personal experience before an event occurs, and then the current operating mode of the power system is actively changed. Then, the dispatcher usually directly issues commands to the event analysis system, thus triggering a corresponding event. In passive dispatching, dispatchers mainly confirm smart decision-making by configuring the SEMS's behavior patterns, the specific configurations of which include the operation mode (open/closed loop), types and triggers of events, event-dispatching strategy, and constraints of smart decision-making.

Dispatchers, who obtain their knowledge and experience from years of education and operation practices, are living "expert systems." The "expert system" formed by dispatchers' wisdom has the highest decision-making authority in the dispatching system, both now and in the future. For example, airliners are equipped with advanced and perfect autopilot systems, and the captain serves only as a supervisor without monitoring when the flying state is set to "automatic"; however, the captain has the right to keep the airliner in a flying state, and then all the control orders are executed by the captain.

8.4 The event analysis model in the SEMS

It was noted in Chapter 2 that this book discusses only three types of indices with decisive significance for power system operation: the security and stability index I_1, power quality index I_2, and low energy consumption index I_3. Correspondingly, three categories of events (i.e., security and stability events, power quality events, and low-energy events) are discussed in this book. Regarding security and stability events, we focus on voltage security events and active power security events. Regarding power quality events, we introduce node voltage quality events mainly. Regarding low-energy events, we introduce network loss economic events and generation cost economic events, mainly. These five events are the most common and significant events with the highest frequency of occurrence.

This book discusses five types of analysis models. The models can be used to determine whether there is an event and, if so, what kind of event is happening. Section 8.5 discusses the processing model for these events (i.e., how to eliminate these events). The following is a brief introduction.

8.4.1 Evaluation of security and stability events

8.4.1.1 Voltage security event

It is assumed that V_r, Q_r are the voltage and reactive power, respectively, of key node r. When the distance from the real-time status point of key node r (V_r, Q_r) to the boundary of the voltage stability region is less than the given threshold, a voltage security event occurs. The expression of this event is

$$(V_r, Q_r) \notin VS_r \tag{8.1}$$

where VS_r is the voltage stability region of key node r and can be described by the shortest distance between it and the boundary of the voltage stability region. The calculation method was discussed in Chapter 5.

8.4.1.2 Active power security event

When the power flows of lines are overloaded, an active power security event occurs. This event is expressed as:

$$P_{ij} > P_{ij}^{\max} \tag{8.2}$$

where P_{ij} is the power flow of the overloaded circuit and $P_{ij}^{\max}$ is the limit of the circuit.

8.4.2 Evaluation of power quality events

When the node voltage is not within the limit, a voltage quality event occurs at the node and is expressed as:

$$V_r \notin [V_{rl}, V_{rh}] \tag{8.3}$$

where V_r is the amplitude of the voltage of key node r, and V_{rl} and V_{rh} are the lower and upper limits, respectively, of the node voltage.

8.4.3 Evaluation of economic operation events

8.4.3.1 Network loss economic event

When the gap between the real network loss of the system and the minimum network loss, calculated by the optimal power flow, exceeds the given threshold, a network loss economic event occurs. The expression is

$$\mathrm{Loss}_{real} - \mathrm{Loss}_{opt} > \varepsilon_{loss} \tag{8.4}$$

where Loss_{real} is the real network loss; Loss_{opt} is the network loss produced by the optimal power flow, the objective of which is the minimum network loss; and ε_{loss} is the threshold, which is usually given based on the operation experience of the dispatchers.

8.4.3.2 Generation cost economic event

When the gap between the system's real (actual) generation cost and the minimum cost that the optimal power flow calculation can achieve (when the difference between the actual cost of the generation of the system and the minimum cost that can be achieved by the optimal power flow calculation) exceeds the threshold, a generation cost economic event occurs. The expression of this event is:

$$F_{real} - F_{opt} > \varepsilon_f \tag{8.5}$$

where F_{real} is the real generation cost; F_{opt} is the calculation result of the optimal power flow, the objective of which is the minimum generation cost; and ε_f is the threshold, which is usually given based on the operation experience of the dispatchers.

8.5 The event processing model in the SEMS

Section 8.4 discusses the analysis model for five types of events: voltage security events, active power security events, node voltage quality events, network loss economic events, and generation cost economic events; the present section discusses the handling model for these events. The role of the event processing model is as follows: to complete the ε transformation, based on the obtained true state set $\mathbf{X}_{TS}(t)$, for sets of different types of events $E_Z(Z=I,II,III)$, and to transform E_Z into the corresponding control command C_Z using a reasonable, effective algorithm.

Instead of dispatching active power and reactive power separately in traditional dispatching, the SEMS emphasizes, comprehensively and collaboratively, exploiting all the controllable resources to achieve the optimal dispatching results. Therefore, it is necessary to solve a variety of OPF problems. A brief introduction is given in the following.

8.5.1 Security and stability event processing

8.5.1.1 Voltage security event

It was noted in Section 8.4 that when the distance from the real-time status point of the key node r to the boundary of the voltage stability region is less than the given threshold, a voltage security event occurs. Then, it is necessary to adjust the system's conditions so that the distance from the real-time status point of the key node r to the boundary of the voltage stability region is maximized after the adjustment. Thus, an optimization model is developed as follows:

$$\begin{aligned}
\max\ & f(V_{pv,r}, Q_{pv,r}) \\
s.t.\ & g(\mathbf{x}, \mathbf{V}_{pv}, \mathbf{C}, \mathbf{L}) = 0 \\
& V_{i,\min} \le V_i \le V_{i,\max}, \quad i = 1, 2, \cdots, n_{bus} \\
& P_{l,\min} \le P_l \le P_{l,\max}, \quad l = 1, 2, \cdots, n_{branch} \\
& Q_{Gi,\min} \le Q_{Gi} \le Q_{Gi,\max}, \quad i = 1, 2, \cdots, n_g \\
& P_{Gi,\min} \le P_{Gi} \le P_{Gi,\max}, \quad i = 1, 2, \cdots, n_g
\end{aligned} \tag{8.6}$$

where n_g is the number of generation units; n_{bus} is the number of nodes; n_{branch} is the number of branches; n_c is the number of capacitors; n_L is the number of reactors; the voltage vector of generation units is $\mathbf{V}_{pv} = \left[V_{pv,1}, V_{pv,2}, \cdots, V_{pv,ng}\right]$; the switching vectors of capacitors $\mathbf{C} = [C_1, C_2, \cdots, C_{nc}]$ and reactors $\mathbf{L} = \left[L_1, L_2, \cdots, L_{N_l}\right]$ are controllable variables; x denotes system status, which is a vector that consists of the voltage and phase angle of each node; $g(\mathbf{x}, \mathbf{V}_{pv}, \mathbf{C}, \mathbf{L}) = 0$ is the power flow equation; $V_{i,\min} \le V_i \le V_{i,\max}$ is the node voltage constraint equation; $P_{l,\min} \le P_l \le P_{l,\max}$ is the constraint equation of branch power flow; $Q_{Gi,\min} \le Q_{Gi} \le Q_{Gi,\max}$ is the constraint equation of the upper and lower bounds of reactive power for the generation units; $P_{Gi,\min} \le P_{Gi} \le P_{Gi,\max}$ is the constraint equation of the upper and lower bounds of active power for the generation units; $\mathbf{V}_{pv,r}, Q_{pv,r}$ are the voltage and reactive power, respectively, of key node r; and $f(V_{pv,r}, Q_{pv,r})$ is the distance from the real-time status point of key node r to the boundary of the voltage stability region. The objective function maximizes the distance from the real-time status point of key node r to the boundary of the voltage stability region after the adjustment. The constraints include the power flow equation, nodes' voltage constraints, branches' power flow constraints, and generation units' active and reactive power constraints.

8.5.1.2 Active power security event

An active power security event occurs when the power flow of branches exceeds the boundary of the power flow. It is necessary to avoid many generation unit adjustments when trying to adjust the power flow of the grid and eliminate this event. Therefore, the corresponding objective function of the OPF model consists of two parts: (1) the range of adjustment for the generation units and (2) the extent that the limits are exceeded. The detailed model is as follows:

$$\begin{aligned} \min \quad & \sum_{i=1}^{ng} (P_{gi} - P_{gi}^{0})^2 + \sum_{i=1}^{n_{ol}} (P_{li} - P_{li}^{\max})^2 \\ s.t. \quad & g(\mathbf{x}, \mathbf{P}_g) = 0 \\ & V_{i,\min} \le V_i \le V_{i,\max}, \qquad i = 1, 2, \cdots, n_{bus} \\ & P_{l,\min} \le P_l \le P_{l,\max}, \qquad l = 1, 2, \cdots, n_{branch} \\ & Q_{Gi,\min} \le Q_{Gi} \le Q_{Gi,\max}, \quad i = 1, 2, \cdots, n_g \\ & P_{Gi,\min} \le P_{Gi} \le P_{Gi,\max}, \quad i = 1, 2, \cdots, n_g \end{aligned} \tag{8.7}$$

where n_{ol} is the number of branches exceeding the limits; n_g is the number of generation units; n_{bus} is the number of nodes; n_{branch} is the number of branches; P_{li} is the active power flow for branch i; $P_{li}^{\max}$ is the active power limits for this branch; P_{gi} is the active power output of generation unit i; P_{gi}^{0} is the active power output before the adjustment; x stands for the system status, which is a vector consisting of the voltage and phase angle of each node; $g(\mathbf{x}, \mathbf{P}_g) = 0$ is the power flow equation; $V_{i,\min} \le V_i \le V_{i,\max}$ is the node voltage constraint equation; $P_{l,\min} \le P_l \le P_{l,\max}$ is the constraint equation of branch power flow; $Q_{Gi,\min} \le Q_{Gi} \le Q_{Gi,\max}$ is the constraint equation of the upper and lower bounds of reactive power for the generation units; and $P_{Gi,\min} \le P_{Gi} \le P_{Gi,\max}$ is the constraint equation of the upper and lower bounds of active power for the generation units. The objective function is to eliminate the active power flow exceeding that of the branch and to minimize the adjustment amount. The constraints include the power flow equation, node voltage constraints, branch power flow constraints, and active and reactive power constraints of the generation units.

8.5.2 Power quality event processing

In the power system, ensuring node voltage quality is an important means of improving the quality of system operation. However, a node voltage may exceed the upper and lower limits in actual operation, in which case, it is necessary to adjust. The goals of the adjustment are to ensure that all the nodes' voltages stay within the upper and lower limits and to minimize the amount of adjustment of the generator terminal voltage, capacitance, and reactance. The OPF model is as follows:

$$\begin{aligned} \min \quad & \sum_{i=1}^{ng}(V_{pv,i}-V_{pv,i}^{0})^{2}+\sum_{j=1}^{nc}(C_j-C_j^{0})^{2}+\sum_{k=1}^{n_l}(L_k-L_k^{0})^{2} \\ s.t. \quad & g(\mathbf{x},\mathbf{V}_{pv},\mathbf{C},\mathbf{L})=0 \\ & V_{i,\min}\le V_i\le V_{i,\max}, \quad i=1,2,\cdots,n_{bus} \\ & P_{l,\min}\le P_l\le P_{l,\max}, \quad l=1,2,\cdots,n_{branch} \\ & Q_{Gi,\min}\le Q_{Gi}\le Q_{Gi,\max}, \quad i=1,2,\cdots,n_g \\ & P_{Gi,\min}\le P_{Gi}\le P_{Gi,\max}, \quad i=1,2,\cdots,n_g \end{aligned} \tag{8.8}$$

where n_g is the number of generation units; n_{bus} is the number of nodes; n_{branch} is the number of branches; n_c is the number of reactors; n_L is the number of resistors; the generation unit voltage vector is $\mathbf{V}_{pv}=\left[V_{pv,1},V_{pv,2},\cdots,V_{pv,ng}\right]$; the switching vectors of capacitors $\mathbf{C}=[C_1,C_2,\cdots,C_{nc}]$ and reactors $\mathbf{L}=\left[L_1,L_2,\cdots,L_{n_l}\right]$ are controllable variables; $V_{pv,i}^{0}$ is the voltage of generation unit i before adjustment; C_i^0 and L_i^0 are the statuses of capacity group i and reactor i, respectively, before adjustment; x is the system status, which is a vector consisting of each node's voltage and phase angle; $g(\mathbf{x},\mathbf{V}_{pv},\mathbf{C},\mathbf{L})=0$ is the power flow equation; $V_{i,\min}\le V_i\le V_{i,\max}$ is the node voltage constraint equation; $P_{l,\min}\le P_l\le P_{l,\max}$ is the constraint equation of branch power flow; $Q_{Gi,\min}\le Q_{Gi}\le Q_{Gi,\max}$ is the constraint equation of the upper and lower bounds of reactive power for the generation units; and $P_{Gi,\min}\le P_{Gi}\le P_{Gi,\max}$ is the constraint equation of the upper and lower bounds of active power for the generation units. The objective function is to minimize the adjustment amount. The constraints include the power flow equation, nodes' voltage constraints, branches' power flow constraints, and generation units' active and reactive power constraints.

8.5.3 Economic event processing

8.5.3.1 Network loss economic event

In terms of the economy, the objective of reactive power control is to reduce network loss, mainly through the rational allocation of reactive power sources in the system (dynamic reactive power devices and static reactive power devices, including capacitive and reactance devices). Therefore, the objective function of the established OPF model also minimizes the system losses. The model is given as follows:

$$\begin{aligned} \min\ & P_{\text{loss}}(\mathbf{V}_{pv},\mathbf{C},\mathbf{L}) \\ s.t.\ & g(\mathbf{x},\mathbf{V}_{pv},\mathbf{C},\mathbf{L})=0 \\ & V_{i,\,\min}\le V_i\le V_{i,\,\max}, \quad i=1,2,\cdots,n_{bus} \\ & P_{l,\,\min}\le P_l\le P_{l,\,\max}, \quad l=1,2,\cdots,n_{branch} \\ & Q_{Gi,\,\min}\le Q_{Gi}\le Q_{Gi,\,\max}, \quad i=1,2,\cdots,n_g \\ & P_{Gi,\,\min}\le P_{Gi}\le P_{Gi,\,\max}, \quad i=1,2,\cdots,n_g \end{aligned} \tag{8.9}$$

where n_g is the number of generation units; n_{bus} is the number of nodes; n_{branch} is the number of branches; n_c is the number of capacitors; n_L is the number of reactors; the vector of generation unit voltages is $\mathbf{V}_{pv}=\left[V_{pv,1},V_{pv,2},\cdots,V_{pv,n_g}\right]$; the switching vectors of capacitors $\mathbf{C}=[C_1,C_2,\cdots,C_{nc}]$ and reactors $\mathbf{L}=\left[L_1,L_2,\cdots,L_{n_l}\right]$ are controllable variables; x is the system status, which is a vector consisting of each node's voltage and phase angle; $g(\mathbf{x},\mathbf{V}_{pv},\mathbf{C},\mathbf{L})=0$ is the power flow equation; $V_{i,\,\min}\le V_i\le V_{i,\,\max}$ is the node voltage constraint equation; $P_{l,\,\min}\le P_l\le P_{l,\,\max}$ is the constraint equation of branch power flow; $Q_{Gi,\,\min}\le Q_{Gi}\le Q_{Gi,\,\max}$ is the constraint equation of the upper and lower bounds of reactive power for the generation units; and $P_{Gi,\,\min}\le P_{Gi}\le P_{Gi,\,\max}$ is the constraint equation of the upper and lower bounds of active power for the generation units. The objective function is to minimize network losses. The constraints include the power flow equation, node voltage constraints, branch power flow constraints, and active and reactive power constraints of the generation units.

8.5.3.2 Generation cost economic event

In the active power control of power systems, the economic requirements rely mainly on the generation costs. The corresponding OPF model is as follows:

$$\begin{aligned} \min\ & \sum_{i=1}^{n_g} f_i(P_{gi}) \\ s.t.\ & g(\mathbf{x},\mathbf{P}_g)=0 \\ & V_{i,\,\min}\le V_i\le V_{i,\,\max}, \quad i=1,2,\cdots,n_{bus} \\ & P_{l,\,\min}\le P_l\le P_{l,\,\max}, \quad l=1,2,\cdots,n_{branch} \\ & Q_{Gi,\,\min}\le Q_{Gi}\le Q_{Gi,\,\max}, \quad i=1,2,\cdots,n_g \\ & P_{Gi,\,\min}\le P_{Gi}\le P_{Gi,\,\max}, \quad i=1,2,\cdots,n_g \end{aligned} \tag{8.10}$$

where n_g is the number of generation units; n_{bus} is the number of nodes; n_{branch} is the number of branches; P_{gi} is the active power output of generation unit i; x is the system status, which is a vector consisting of each node's voltage and phase angle; $g(\mathbf{x}, \mathbf{P}_g) = 0$ is the power flow equation; $V_{i,\min} \le V_i \le V_{i,\max}$ is the node voltage constraint equation; $P_{l,\min} \le P_l \le P_{l,\max}$ is the constraint equation of branch power flow; $Q_{Gi,\min} \le Q_{Gi} \le Q_{Gi,\max}$ is the constraint equation of the upper and lower bounds of reactive power for the generation units; $P_{Gi,\min} \le P_{Gi} \le P_{Gi,\max}$ is the constraint equation of the upper and lower bounds of active power for the generation units; and $f_i(P_{gi})$ is the generation cost of generation unit i when the output is P_{gi}. The objective function is to minimize the generating cost of the whole system. The constraints include the power flow equation, nodes' voltage constraints, branches' power flow constraints, and generation units' active and reactive power constraints.

8.6 Controllable resources of the SEMS

The goal of the SEMS is to realize multi-index optimality-approximating operation; thus, it is necessary to regulate all the controllable resources in the system under the guidance of power system hybrid control theory.

What are the controllable resources? Strictly speaking, all possible means of changing the system operation status can be treated as controllable resources. Then, the use of controllable resources should be unified, coordinated, and made optimal.

What are the controllable resources within the system? Narrowly speaking, the regulation of the power system is realized by reasonable regulation of various types of power system equipment, and this approach can ensure the security, stability, quality, and economy of the power system. In this sense, the controllable resources include the active and reactive power of generators, FACTS equipment of substations, static volt-ampere (var) generator (SVG), SVC, capacitors, and transformer taps. Additionally, a power load with demand-side response ability, the breaking data of multiple voltage levels, the "opening and closing" of switching devices, etc., can be treated as controllable resources. Generally, the regulation of the power system is realized by rational planning and allocation of electrical power equipment. In this sense, the controllable resources of the power system include not only the above-described resources in a narrow sense but also the planning and construction of power equipment and primary energy (such as wind, hydro, natural gas, and coal resources).

The differences in action time between different controllable resources are very large. For example, a circuit breaker may achieve operation within 0.1 s, but starting up a thermal generation unit takes hours. Furthermore, it takes several months to achieve annual/seasonal regulation of the reservoir storage capacity of hydropower stations. Therefore, regulation at different time scales requires the implementations of different resources. For example, in real-time dispatching, the main consideration is the active and reactive power of generation units. However, in annual dispatching, the main consideration is the annual/seasonal regulation of the reservoir storage capacity of hydropower stations. Figure 8.5 illustrates the response times of common controllable resources.

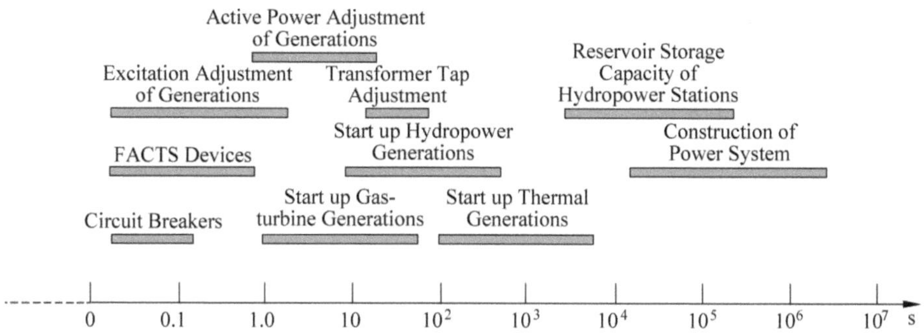

Figure 8.5: Actuation time with different control methods in the power system.

The controllable resources in the system are numerous. They are introduced by the following classifications.

8.6.1 Classification by the information utilized in system control

Power system control can be classified by the information used in system control into global control and local control.

Local control means that devices are controlled by using local information. The advantages are that the design is simple and the regulations is quick. A disadvantage is the lack of coordination between interconnected systems; thus, it is difficult to ensure global optimization. The regulations that belong to local control in power systems include power system protection, regional automatic generation control (AGC), generator excitation control, and AGC- and AVC-coordinated regulation of local units mainly.

Global control means that the power grid is controlled by using the global information of the power system. This type of control is generally located in the power control center. The remote control and adjustment of power equipment are achieved by SCADA/WAMS. The main control methods belonging to global control are remote AGC and AVC, cutting machines, load shedding, emergency backup power support, etc.

8.6.2 Classification by the response time of system control

Power system control can be classified by response time into two types: real-time control and ahead-of-time control (or advanced control and anticipatory control).

(1) Real-time control is based on real-time information, and the required response speed of the control equipment is usually several seconds.
(2) Ahead-of-time control uses prediction results for control devices, with an "advance amount" of time ahead when the operation time of the controlled devices is relatively long. Thus, unfavorable situations caused by control measures' lagging behind the development of the power system state can be avoided.

8.6.3 Classification by power system operation status

Power system control can be classified by power grid operation status into preventive control, emergency control, and recovery control.
(1) The stable operation of the power system requires maintaining a certain safety margin. When the safety margin is insufficient, there are risks even when the power system is operated under normal conditions. The measures taken to improve the power system stability margin can be called preventive control. As the above concept shows, preventive control does not require a high response speed of regulation; thus, power flow adjustment can be used.
(2) After a disturbance in the system, some control measures should be performed to keep the system operating in a viable operation state and avoid unacceptable overload or abnormal frequency (or voltage). This approach is called emergency control. Compared with that under normal circumstances, the response time requirement for emergency control is very stringent. Cutting machines, load shedding, and even emergency splitting may be used, where necessary to ensure the stability of the system.
(3) Recovery control means a control process to restore the system to the new operating state in the shortest possible time when a failure, partial load power outage, splitting the state, or even a whole network outage happens in the power system. Therefore, we must develop recovery manipulation plans in advance and implement them promptly and accurately after incidents. Doing so is the only way to minimize economic losses caused by large-scale power system blackouts.

8.7 The layered hierarchical structure of the SEMS

To coordinate with a layered dispatching system, an appropriate SEMS should be developed at each level of scheduling, thereby forming a geographically dispersed, layered hierarchical smart dispatching automation system.

n the layered hierarchical control system, the general structure of one level of the system is shown in Figure 8.6. At this level, the system consists of the SEMS and devices that are controlled by the SEMS, such as stations, plants, FACTS devices, distributed generation, power load, and lower-level SEMSs. Additionally, the system is required to receive control commands from upper-level SEMSs, and exchange data and information with SEMSs at the same level. In addition, the system needs to exchange data with external systems (e.g., a marketing system) at this level. This information exchange is realized through the data-sharing platform.

It is obvious that the responsibilities of the SEMS should include the following three parts:

(1) The upper-level SEMSs works (function) as an intermediate layer to execute control commands from the upper-level dispatching.
(2) For stations and plants controlled by the SEMSs at this level and lower levels, according to the specific situations, the system can function as a decision-making level to issue control commands or at an intermediate level, to issue manipulation commands.
(3) Measurement data and other information are exchanged between the SEMS and other application systems at the same level.

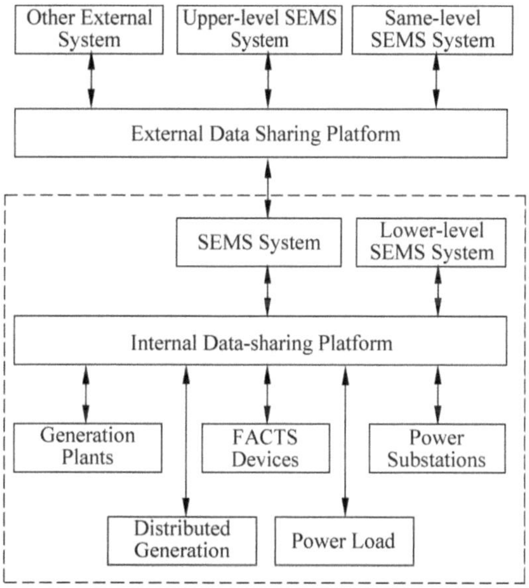

Figure 8.6: Layered hierarchical structure of the SEMS.

8.8 Conclusions

In this chapter, the definition, characteristics, components, and structures of the SEMS are introduced. The event analysis model in the SEMS and the corresponding event processing logic are introduced.

In summary, the SEMS is an engineering implementation of an SPS in dispatching control. The SEMS is integrated by using key SPS technologies, such as hybrid control methods, standards-based system operation indexes, data analysis, and optimized control algorithms. Therefore, when a large power system is upgraded to an SPS, the SEMS of dispatch centers (or stations) at all levels must first be established. The power system technology innovation practices in any country, region, or community will verify this assertion. This approach is the only way to realize the future DPS.

Chapter 9
Smart grid

9.1 Background

Distribution networks are one of the essential components in the modern power system for the following reasons: 1) Power companies, including State Grid and China Southern Power Grid Company, provide energy services to more than one billion customers, which are connected to the distribution network. 2) Since the distribution grid is at the end of the power system network, the total number of transformers and the sum of the lengths of lines in the distribution grid are much larger than those in the transmission network. Approximately 60% – 70% of the network losses occur in the distribution grid. As a result, there is high potential for reduction of energy loss. 3) Many distributed renewable energy resources such as hydro, wind, solar and electric vehicles are located in distribution networks. 4) Most importantly, daily life of residential users highly depends on reliable power supplies from distribution grids.

Thus, most of the demands and losses occur in distribution networks. The applications of new renewable energy resource technologies and new techniques typically start from distribution grids. Therefore, there is an acute need to improve the monitoring of these distribution grids and enhance the management and operation qualities by applying advanced technologies to achieve a secure, high-quality, and reliable power system. Achieving such a system has become the goal not only for power companies but also for national development. To achieve this goal, we must significantly reform distribution grids and apply almost all the advanced technologies related to power systems from information technology (IT), industrial control, and manufacturing areas to construct a distribution-side dispatch-smart energy management system (D-SEMS) so that distribution networks will evolve into real "smart grids." Therefore, in this chapter, we discuss in detail what a smart grid (SG) is.

9.2 Definition of a smart grid

A distribution network is a key part of a power system. However, since there are no large-scale generation units or high-voltage direct-current (HVDC) transmission lines in this part, this network is purely an alternating current (AC) network. Above all, a distribution network is still a grid and a part of the power system; thus, the definition of an SG should be consistent with that of a smart power system (SPS).

https://doi.org/10.1515/9783110448825-009

Definition 9.1 An SG is an SPS at a 110 kV voltage level or below.

Definition 9.1 is concise but too brief for readers to understand. Thus, by inserting the definition of an SPS in Section 1.2.1 into definition 9.1, a more detailed definition of an SG is given as follows.

Definition 9.2 An SG is an SPS at a 110 kV voltage level or below and is capable of multi-index optimality-approximating operation.

The optimization results depend on the selected metrics of interest. Then, what are the similarities and differences between the optimizing indices in SG operation and the multiple indices in Section 1.2.2? Specifically, both groups of indices are consistent: both cover three primary aspects – I_1 (security indices), I_2 (quality indices), and I_3 (loss indices). However, compared to the multiple indices in an SPS, these three indices for an SG are distinctive in terms of specific contents.

First, to meet the essential need to provide electricity to the demand side, an SG should assign the highest priority to reliability in supplying electricity to customers. We can connect the reliability metric improvement to a decrease in the average time of loss of load. In other words, when the average interruption hours of customers are minimized, the optimal security indices in an SG are achieved.

Second, as mentioned in previous chapters, most losses are from distribution grids. Thus, a reduction in the network loss can also increase the economic metrics of an SG. In addition, since there are many small-scale distributed renewable resources in the distribution network, the effective utilization of those units will definitely improve the efficiency of the power system and reduce the emissions of polluting gases. Therefore, the economic metrics of the SG include both loss reduction and emission reduction.

Third, the improvement of electricity quality is an important operational index for optimal SG operations. On the one hand, the qualified rate of the voltage amplitudes should increase (the rate should be 100%), and harmonics in the grids should be minimized. On the other hand, avoiding sudden voltage drops and guaranteeing a balanced energy supply are also important in maximizing the quality of the supplied power.

To achieve the optimal operations of an SG through the three aspects mentioned above, we need to reform the distribution network infrastructure, improve the dispatch and control strategies, and apply advanced techniques so that the distribution network evolves into an SG.

9.3 Construction of a modernized distribution grid

Currently, the automation level in dispatch and control processes for distribution networks is relatively low, and these networks are far from fulfilling the objectives of security, reliability, and economics required by an SG. We would like to note several reasons: (1) There is a misunderstanding that only high-voltage, large-scale networks require an energy management system (EMS) and that low-voltage distribution networks do not. (2) Most of the researchers and management staff in power companies graduated from colleges and not from power-focused universities. Since these persons pay less attention to distribution networks, they have limited interest in distribution network development when they are employed full-time at power companies after graduation. As a result, this situation is not conducive to improving the standards of distribution grids. (3) Many distribution grids are located in rural areas and have received less attention, but this situation has largely changed in recent years. (4) The ideas that people are trapped in ingrained attitudes, that is, voltage regulation, frequency responses, and load characteristics depend primarily on large-scale units and hydro plants. Using major thermal power plants for peak load regulation is not an innovative approach. It is even worse when thermal units are required for deep peak load regulation, thus increasing their emissions and lowering their operational efficiency. This approach is purely a temporary solution that is considered only at the very beginning of distribution network development. Once all the loads can actively participate in peak load regulation, there is no need to adopt such an ineffective approach that uses only these thermal power plants. (5) Since distribution networks cover almost all cities, towns, and rural areas, a very large investment in the long-term development of distribution networks is required. Therefore, there is an acute need for educating people with correct, up-to-date concepts and knowledge. This need is also one of the purposes of writing this book. (6) Shortcomings in implementing advanced communication and IT techniques limit the application of a flexible real-time pricing mechanism and impede the interactivities between the demand side and the power grid. Therefore, to construct a mature SG, we need to invest in education, public policies, technical mechanisms, and technology aspects. First, we need to change minds by introducing advanced dispatch and management methods, improving technical mechanisms, redesigning regulation policies, and enforcing demand-side management. Then, we must construct a D-SEMS for the distribution side and foster the interactivities between the demand side and the power grid to enhance the benefits of considering load characteristics in dispatch decision-making procedures. Finally, we should update D-SEMS-related theories, techniques, and approaches to improve the closed-loop control capability of multi-index optimization in power system operations.

9.4 Demand-side management for peak load regulation

The approach and mechanism proposed in Section 9.3 enable the interactivities between the demand side and power system operation and dispatch procedures, thereby guaranteeing that the optimization of the power system load curve can still be achieved even if all major thermal units are disabled for peak load regulation.

In modern society, electricity is one of the most fundamental productions and living consumptions. It is better for each customer to be informed of the real-time prices when they are consuming electrical energy. Thus, electricity price significantly affects industrial and residential activities. As electricity prices change over time, if conditions permit, most industrial, commercial, and residential customers can make active choices to use electricity during low-price periods or to deliver and sell their stored energy when the electricity price is high. For example, cash-back cars and power generation systems based on air energy storage are operated similarly. There is a natural supply-demand relationship in an active electricity market. Therefore, determining appropriate dynamic electricity prices for different customers and virtually showing the price curves to the customers will help guide the behaviors of these customers and enable comprehensive regulation of the load curves.

Support from both technology and infrastructure is necessary to adjust the whole system load dynamic characteristics by setting electricity prices. First, we need to apply advanced IT techniques to economic activities in distribution networks by designing, constructing, and implementing advanced metering infrastructure (AMI) to monitor demand-side activities in real-time. Then, if we can dynamically publish the electricity prices for the upcoming hours through the Internet, cable, and the broadcast media, customers can actively adjust their energy consumption to achieve a desired power system operation status. In addition, the system operator can provide the most economical dispatch strategy, based on the importance and characteristics of different loads. For example, during high-demand summer periods, setting different price curves and providing various advices for residential and agricultural electricity consumption can be done to achieve peak shaving and valley filling. Ultimately, the gradual construction of a competitive electricity market can encourage customers to participate in electricity production, and customers can benefit from trading activities.

Based on the above analysis, in future power systems, the load will become the main aspect in a load characteristic curve control process. We basically need to solve the most important problem – the differences between peaks and valleys. The solution to this problem will bring great economic and social benefits. Solving this problem is the pathway to green development in China.

9.5 Two-sided energy management systems for SGs

The most distinctive difference between an SG and an SPS with a 110 kV voltage level or above is that an SG directly serves the demand side: all industrial, commercial, and residential customers. An SG should only provide – not enforce – correct guidance and suggestions. However, the situation is different in an SPS. For example, automatic generation control (AGC) and automatic voltage control (AVC) commands sent from a smart energy management system (SEMS) at the system control center to reset the relay protection setting values cannot be changed by any staff member of either a power plant or a substation; else, the whole power system might be operating in disorder. This restriction results in the challenge of effectively managing the demand side compared to energy management in high-voltage large-scale power systems. This restriction is also why there has been limited improvement in energy management on the demand side, even though such improvement has been proposed for years. In summary, to achieve multi-index optimal demand-side energy management, we need to build two-sided SEMSs for distribution networks: a user-level smart energy management system (U-SEMS) for demand-side energy management and a distribution-level smart energy management system (D-SEMS) for local dispatch centers.

9.5.1 User-Smart Energy Management System (U-SEMS)

As mentioned above, the most significant benefit from developing an SG is building an appropriate pricing mechanism to encourage customers to participate in peak load regulation as the major component. To do so, we need to build a U-SEMS with the following functions:

1. Data acquisition function
U-SEMS has two means of data acquisition: the U-SEMS can install meters and use smart domestic appliances to read real-time data – real power, reactive power, voltage, power factor, and so on; the communication system can exchange information – real-time price, frequency, harmonics, and so on – with the power system control center.

2. Energy management function
Energy management refers to users being able to adjust their electricity consumption activities according to appliance statuses, priority lists, needs for energy, and so on. For instance, when the real-time electricity price is high, we need to guarantee energy supply to those appliances with highly strict requirements for minimum on/off time or high priorities. During low-price periods, users can turn on appliances with high energy consumption, such as heaters and dryers. This operation strategy is similar to that used in agricultural activities. The storage devices and distributed energy resources should also be considered and managed in the energy management function, if installed.

3. Data analysis function
The data analysis function can provide users with a detailed report on their historical electricity consumption activities and corresponding advice for future improvements.

4. Remote control and adjustment function
This function enables customers to control their appliances and adjust their energy consumption through an internal communication network. Consequently, customers can adjust their total electricity consumption according to the power system states.

5. Remote management function

When customers are not at a location, this function enables customers to remotely monitor their energy usage status, so that electricity consumption activities can be adjusted and a few unnecessary appliances can be turned off.

6. Advanced functions

The U-SEMS can also install sensor modules to detect human activities and feelings. For example, when sensors detect no one in a room, the system can automatically turn down or turn off some appliances. By collecting the historical electricity usage data of customers, the U-SEMS can also build a specified energy consumption model and optimize the model through techniques such as generating a set of advices to improve energy consumption and save energy.

In addition, these collected data can be uploaded to the system control center for further utilization, such as the characterization of local energy consumption. These analysis results can help in building an optimal plan for network planning, electricity pricing, and equipment maintenance schedules. Thus, we can improve network efficiency, minimize loss, and so on.

9.5.2 Distribution-Smart Energy Management System

Once we notice the importance of distribution network improvement for maximizing social welfare, we will agree that an energy management system (EMS) is useful not only in high-voltage grids but also in low-voltage distribution networks. The same is true for both urban and rural areas.

The construction of a D-SEMS still aims to enable the distribution network to achieve multi-index operational optimality. Similar to a SEMS, a D-SEMS also has three operational metrics – I_1 (security indices), I_2 (quality indices), and I_3 (loss indices).

High security refers to a reduction in the average outage time of customers. Specifically, for the most important customers, we need to utilize specific control, self-recovery, and restructuring methods to guarantee that the outage time is close to zero, in one year.

High electricity quality means that the voltage levels, sudden voltage drops, and voltage harmonics at each node meet the national and industrial requirements.

Network loss minimization helps us operate power systems in an environmentally friendly way. Ten percent of the total produced energy is consumed via network loss, of which 60% ~ 70% occurs in a low-voltage distribution network (such as the driver devices for large or medium-sized motors with no frequency control modules). If we operate the distribution network through the D-SEMS, we can obtain the optimal power flow solution for the distribution network and smart reactive power compensation to lower the loss by approximately 2%. Therefore, we can save 200 billion kWh per year, which is not negligible.

Specifically, the three metrics of interest are consistent for both power systems and distribution networks. However, due to the very large differences between large-scale systems and low-voltage distribution networks, there are various standards for each metric. The theories, approaches, network structures, and ideas for constructing D-SEMSs like smart wide area robots (S-WARs) are the same as the descriptions in Chapter 2.

As mentioned in the above sections, it is more complicated to build SEMSs for distribution networks than for large-scale systems. Thus, to construct the D-SEMS, a few new techniques need to be developed with some novel characteristics.

9.6 Implementation of new techniques for the development of SGs

To obtain multi-index optimal operations, an SG must utilize advanced computer devices and information systems for replacing manual controls to realize automatic closed-loop control of distribution grids. Thus, we need to keep developing new techniques, new approaches, and new equipment to enrich the control methods for distribution networks and improve network automation levels.

Currently, there are several new techniques for constructing SGs:

(1) Distributed energy resources (including wind farms, solar plants, tidal energy resources, biogas generation units, and so on) and cooling-heating-electricity cogeneration plants.[1] In the current energy flow system, we can utilize multisite distributed conventional units and renewable energy resources with capacities no greater than 200 MW to form cooling-heating-electricity cogeneration plants. This method provides a new direction for efficiency improvements and emission reductions.

(2) Storage techniques, consisting primarily of mechanical, electrochemical, electromagnetic, and phase-change cells. The advantages of these storage devices include the following: (a) The characteristic curve of system load is improved. This improvement will be important in peak-load shaving. (b) The difficulties in integrating distributed energy resources into the power system are reduced. The application of storage devices can mitigate the uncertain fluctuations caused by renewable energy resources. (c) The reliability of the energy supply to high-priority users is improved; in high-priority cases, equipment and other materials could be harmed with a loss of power even for only a short period. When a power

1 The authorities in some areas mandate that all units with capacities lower than 200 MW be put offline and destroyed to demonstrate their determination to reduce emissions. However, we do not think this is the most effective solution. Instead, we can use these plants to replace heat generation units of equivalent capacities for heat generation during winter periods. Thus, we can actually achieve the low-emission goal.

outage occurs and the emergency energy supply cannot respond in time, the energy storage power sources can be used to supply power quickly.
(3) Economic incentive technology, such as smart appliances, demand-side management, and time-dependent pricing mechanisms. These techniques motivate users to participate in peak-shaving and valley-filling activities, according to time-varying prices.
(4) Innovative control strategies for distribution network management, including virtual generation plants and microgrids: Virtual generation plants combine many distributed energy resources for commitment and dispatch. A microgrid integrates generation units, loads, storage devices, and control modules to construct a controllable unit such that we can supply heating, cooling, or electricity and implement the commands from the power system control center. User monitoring can be strengthened by AMI. Additionally, users can obtain relevant information to adjust the operation status of generation and distribution networks by monitoring; making such adjustments is beneficial to both users and power grids.

In summary, to control a distribution network such as a smart wide area robot and achieve an SG, we must upgrade current electric equipment and communication networks. We also need to widely apply distributed generation units, storage devices, economic incentive resources, and innovative distribution network control strategies to automatic, closed-loop, optimal operations for distribution grids, as described for the D-SEMS and U-SEMS.

9.6.1 Utilization of renewable energy resources

Conventional units use fossil fuels as energy resources for electricity production, thereby causing serious pollution problems. Thus, people are paying increasing attention to the development of renewable energy resources. Compared to conventional generation units, distributed energy resources refer to cooling/heating/power supply systems that are small-scale (several kW to 50 MW) and are sparsely distributed among distribution networks.

Distributed energy resources refer primarily to solar, wind, tidal, and biogas generation units and a few cooling/heating/power cogeneration plants. Storage devices are also a type of distributed energy resource. Here, we need to pay sufficient attention to the massive utilization of storage devices because storage devices are relatively expensive, and the life cycle of typical batteries is approximately 5–6 years. These storage devices cause additional harm to the environment. High-density batteries can also be applied as uninterruptible power supply (UPS) devices for high-priority users such as hospitals and military sites.

(1) Aside from hydroelectric power, wind power is the most important and economical renewable energy source, with an annual incremental installation rate of 24.7% from 2003 to 2007 and a total capacity of 94 GW by 2007. Approximately 1% (200 billion kWh) of the global energy consumption is supplied by wind energy resources [84]. However, wind power is uncertain, variable, and intermittent, with an annual utilization factor of approximately 2,000–2,500 h per year.
(2) Solar power is widely available, clean, and secure, thus making it one of the most promising renewable technologies. However, the energy density of solar irradiation is very low, and the installation costs of solar plants are relatively high. In addition, solar plant performance depends heavily on weather conditions, with no output at night. These characteristics limit the development and implementation of solar plants. However, some areas that are unsuitable for agriculture – areas in Tibet, Xinjiang, and Inner Mongolia and on rooftops in coastal cities of China – are the best places for utilizing solar energy.
(3) Fuel cells are clean and quiet and have high energy conversion efficiency (45% ~ 60%), a high energy density, a short response time, quick start-up, zero emissions, and other advantages. Fuel cells can be used for portable energy supplies, electric vehicle storage components, and so on. Fuel cells can supply energy in both centralized and distributed ways [85, 86].
(4) Tidal energy resources use potential and kinetic energy during tidal activities for electricity production. Biogas generation units convert biomass into biogas for burning and generating energy. Compared to other distributed energy technologies, tidal and biogas energy resources are two new distributed generation techniques with limited application due to their locational dependences. The utilization of biogas is very important for users in rural areas and has resulted in a new research topic. We can use biological enzymes to generate biogas from waste, and a few byproducts can be processed further and used as fertilizer. These benefits are of great interest to researchers in biology and chemistry.
(5) Gas turbine cogeneration units fully use fossil fuels for cooling, heating, and electricity supplies. In addition to the electricity generated, the systems also output waste heat for heating and cooling. This technique is relatively mature and cost-effective and can be used widely.

In the following sections, we give a brief overview of the renewable energy technologies mentioned above.

9.6.1.1 Wind power

Wind power is an environmentally friendly renewable energy source. Abundant wind energy is available and can be harvested for electricity production, across China. Based on the up-to-date survey results published by the State Meteorological Administration, the total available electric energy from wind power is approximately

400 GW, with 250 GW on land and 150 GW from offshore. Reports indicate that the total installed wind power capacity may reach 100 GW by the end of 2020 and 300 GW by the end of 2030. The annual utilization hours will increase from 150 h to 2,500 ~ 3,000 h. Areas with abundant wind power are usually far away from load centers and located in relatively weak power grids. At the end of 2009, the cumulative installed wind capacity was less than 26 GW, supplying less than 1% of the national demand. There are many studies on how to effectively integrate more wind farms into the grid during the construction of SGs and SPSs.

Based on the scales and utilization approaches, there are two types of wind farms: off-grid wind farms and grid-connected wind farms. Off-grid wind farms are not necessarily connected to power grids and typically meet local demands, such as offshore drilling platforms, which are not connected to any power system. For a consistent and reliable power supply, it is better to install solar plants and storage devices to construct a stand-alone microgrid. However, the initial construction of such a microgrid always depends on financial support from the government.

Grid-connected wind farm technology became a major force after 1990 and had two distinctive types: 1) A few grid-connected wind farms are integrated into the power system, and they are connected to local distribution networks with voltage levels lower than 35 kV. These wind farms are typically dispatched by the control center, together with local small-scale hydro plants. Hainan Province is planning to adopt this technique. 2) Other grid-connected wind farms are large-scale and supply electricity to grids through high-voltage transmission networks with voltage levels higher than 110 kV. The capacities of these wind farms are usually n × 100 MW or n × 10 GW. Most areas with abundant wind energy are located in northern or western China and are always far from load centers. The wind farms in Gansu, Inner Mongolia, and Xinjiang Provinces are of this type.

As the details of wind turbine design and manufacturing are outside the scope of this book, we discuss electricity quality issues with the increasing penetration of wind power below.

The n × 1000 MW wind farms installed in Inner Mongolia and/or n × 10 GW wind farms installed in Gansu Province are two of the largest wind energy resources worldwide. The integration issues in these wind farms pose new challenges to the system operators. These challenges are not difficult to solve. However, the development of wind power is increasing too quickly, so we have not had a chance to consider these issues. Figures 9.1 and 9.2 illustrate two different control and operation strategies for power systems with installed wind power. A coordinative control strategy for gas turbine units, wind, and hydropower is shown in Figure 9.1. In addition, Figures 9.2 and 9.3 show operation approaches to mitigate the impacts of wind power uncertainty on the power balance and the system frequency. As shown in Figure 9.2, if there are insufficient gas turbine units and hydro plants installed in power system A, we must have enough gas and hydro power plants in the external grid for regulation.

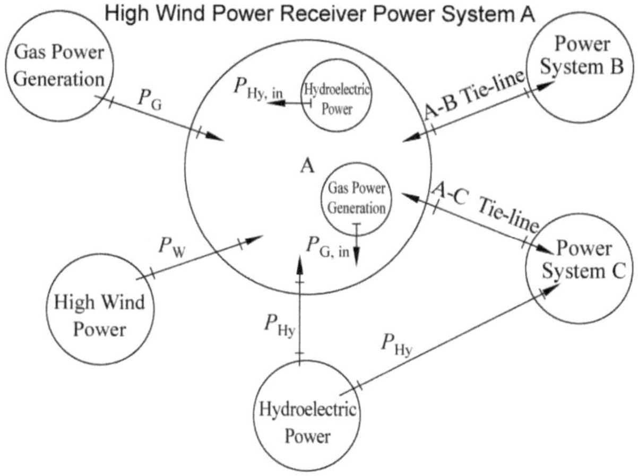

Figure 9.1: The coordinative operations of large-scale wind farms, hydro plants, and gas turbine units.

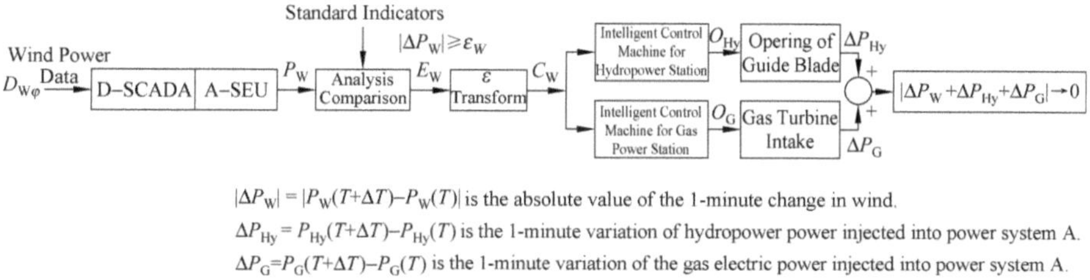

Figure 9.2: Centralized control strategies for a SEMS to manage wind power fluctuations.

Based on the centralized control strategy provided in Figure 9.2, P_G and P_{Hy} are determined by the SEMS of system A. When P_W increases or decreases during the period $[T, T+\Delta T]$, the wind power deviation ΔP_W, the system frequency f, and the rate of change in the system frequency $\left(\frac{df}{dt}\right)$ change correspondingly. We refer to E_W as the contingencies caused by significant wind power fluctuations, namely:

$$E_W = \{E_{\Delta P_W}, E_{\Delta f}, E_f\} \tag{9.1}$$

The command conversion $\varepsilon(\cdot)$ is used to eliminate the contingencies E_W such that the frequency deviation can be maintained within the interval of –0.2% ~ 0.2%.

As shown in Figure 9.3, the operational status information from large-scale wind farms, hydro plants, and gas turbine units is sent to both the SEMS (A) and the information-sharing platform installed in the self-control systems. The SCADA system and state estimator in the self-control systems analyze whether $|P_W|$ is within the threshold values. If not, a contingency signal E_W is triggered, and the corresponding control signal C_W is generated to regulate the output of the hydro and gas turbine units, such that the frequency deviation is maintained within –0.2% ~ 0.2%. However, the internal hydro and gas turbine units are still under the control of the SEMS (A). Thus, the proposed control strategy mitigates the burdens on the SEMS (A).

Both proposed strategies are feasible and require the implementation of wind forecasts, data visualization techniques, and other ancillary services.

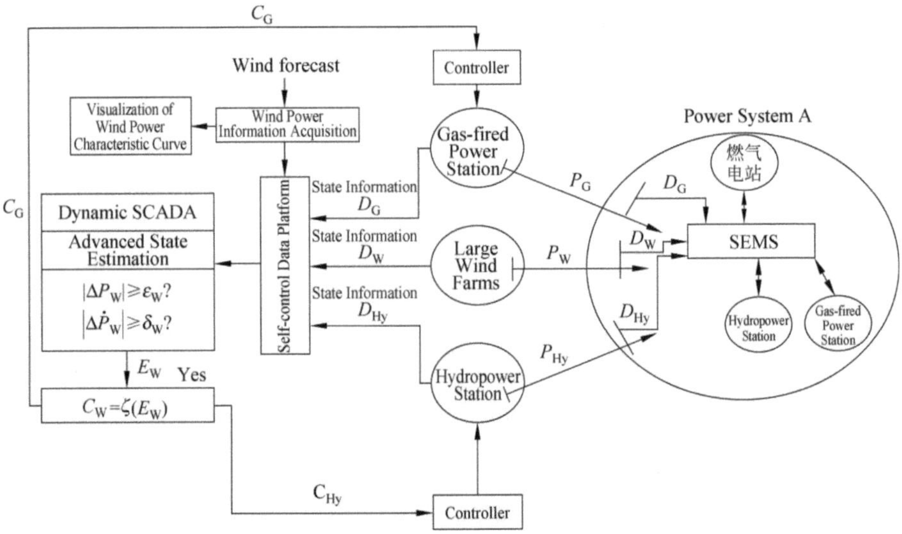

Figure 9.3: Diagram of frequency regulation for power systems with large-scale wind farms.

9.6.1.2 Solar power

The total solar power reaching the Earth's atmosphere is reportedly 173,000 TW, of which 81,000 TW can reach the ground [87]. If we collect all the solar energy reaching the ground for just one hour, the collected energy could supply the yearly demands of all human beings. In addition, solar power plants require relatively low water usage and do not emit any greenhouse gases. Consequently, many countries and districts are urging the development of solar techniques for supplying electricity demands and motivating economic development. A sizable investment from these authorities is used to support solar techniques, such as reducing installation/maintenance costs and improving solar plant performance. Research reported by the European Photovoltaic Industry Association (EPIA) indicates the rapid development of the solar power industry within the last decade, with a strictly increasing annual development rate [88]. From 1998 to 2008, the cumulative installed solar plant capacity increased by 30% annually, on average. In the solar power forecast report provided by the International Energy Association (IEA), 1% of the total energy consumption worldwide is predicted to come from solar plants by the end of 2020, and this will increase to 20% ~ 25% by 2050 [89].

China has vast amounts of solar energy across the country. The total solar energy received by land in China is forecasted to be approximately 14.7 billion GWh.

This value is equal to 4,900 billion tons of standard coal. The annual solar energy received by land at different locations in China varies from 335 ~ 837 kJ/cm^2 with a median value of 586 kJ/cm^2. Based on the solar energy distribution over the land in China, Tibet, Tsinghai, Xingjiang, the southern part of Inner Mongolia, Shanxi, the northern part of Shan'xi, Hebei, Shandong, Liaoning, the western part of Jilin, central and southwest Yunnan, the southeast of Guangdong, the southeast of Fujian, the eastern and western parts of Hainan, the southwest of Taiwan, and some other areas have relatively high solar energy available for harvesting. Specifically, compared to other areas, the plateau area in Tibet has more solar energy due to the area's thin and clear atmosphere, high transparency, low latitude, and long sunrise-to-sunset periods. On the other hand, the total area of wilderness in China is approximately 1.08 million km^2, of which 0.85 million km^2 are desert, desert lands, and semidesert lands. We assume that a 50 MW solar plant requires a 1 km^2 area and that the total installed solar capacity will be 10 GW by 2020; if solar plants totaling 6 GW are sited in the wilderness, an area of only 120 km^2 (0.015% of the total area of the wilderness) is required. Therefore, the natural conditions in China are definitely advantageous for the development of solar energy resources (particularly large-scale grid-connected solar plants). With the implementation of renewable energy rules and a few other related policies, the contribution of solar energy resources has become increasingly significant. In 2010, the incremental solar capacity was higher than 500 MW. From 2011 to 2020, China's cumulative installed photovoltaic capacity has increased from 3GW to 252.5GW. According to the National Energy Administration, as of June 30, 2021, China's cumulative installed photovoltaic capacity has reached 267GW, a surge of 8800% in ten years.

Similar to the categorization of wind farms, we can also classify solar energy resources into two distinct types: off-grid and grid-connected. Off-grid solar plants are used primarily to supply electricity to rural areas, coastal military forces, and streetlights. Based on whether the solar plants are equipped with storage devices, we can divide grid-connected solar plants into dispatchable and nondispatchable solar energy resources. The deviations from forecasts of the power output for nondispatchable solar plants are balanced by the power systems. Dispatchable solar plants can mitigate the variation and intermittency of the output by using installed storage devices. With the development of storage technology, increasing attention has been given to dispatchable solar energy resources.

Currently, there are two major methods of converting solar energy into electricity: the first approach, namely, a solar thermal power plant, converts solar energy into thermal energy, which is used to generate electricity. The second approach, called photovoltaics, utilizes photoelectric cells to convert solar energy directly into electrical energy. Due to the decreasing costs of semiconductor components and the development of photoelectric materials, photovoltaics are widely used in solar plants in the absence of steam generators or turbines.

Photoelectric cells are components that use PN junctions in semiconductors to convert solar energy into electricity. This energy conversion is called the photovoltaic effect. Due to the voltage and capacity limits of a single photovoltaic cell, many photovoltaic cells are connected in parallel and in series to supply energy. Silicon batteries and thin-film batteries are the two primary categories of photovoltaic batteries. Silicon batteries can be further divided into monocrystalline silicon and polycrystalline silicon batteries. Their conversion efficiency is typically approximately 15% (in some, efficiencies are above 25%). However, silicon batteries are relatively expensive to produce, even though they occupy the highest market share. Thin-film photovoltaic batteries have relatively low efficiencies ranging from 5% to 8%. Their production costs are also relatively low compared to those of silicon batteries. In recent years, the conversion efficiencies of silicon batteries, such as for amorphous silicon cells and cadmium telluride cells, have significantly improved. These cells find application in desertified (desert) lands that have sufficient solar energy.

Half of solar cell production costs arise from highly purified silicon. Due to imperfect competition in the silicon market, silicon battery production costs continue to increase with limited future reduction possibilities, and this aspect negatively affects the development of silicon battery technology. Although China is the country with the highest production of silicon batteries, most of the production comes from small-scale, low-efficiency, and highly polluting manufacturers. China has become the primary source of silicon photovoltaic batteries for Europe. Europe is becoming cleaner, but the pollution in China has worsened. Determining how to change this state is a problem worthy of discussion. The key to developing solar plant techniques is to improve the performance of photoelectric materials. Even though thin-film photovoltaic batteries have low efficiencies and short life, the implementation of cryogenic techniques lowers the energy loss in these batteries. The utilization of other inexpensive materials further stabilizes their production costs. Since there are few constraints on large silicon producers, their production costs are easily controlled.

The future development directions and strategies for solar cells are as follows: 1) The first is to develop silicon batteries with higher conversion efficiencies and a longer life (20–25 years). The products are exported mainly to Europe and America to earn foreign exchange and achieve a stable level of employment in China. However, we need to note that this strategy will make Europe cleaner and lead to more pollution in China, without the development of hydro and nuclear plants. 2) The second is to develop amorphous silicon cells with both higher conversion efficiencies and lower pollution. These plants are installed primarily in Inner Mongolia, Xinjiang, Tibet, and a few other areas. 3) The third is to improve the performance of photovoltaic materials with higher conversion efficiencies and lower costs.

9.6.1.3 Fuel cells

Fuel cell technology, proposed by G. R. Grove in 1839, converts chemical energy from the electrochemical reactions between hydrogen/natural gas/coal gas and oxidants into electric energy.

When fuel cells produce electricity, the reactants are sent to the positive and negative electrodes that are separated by an electrolytic liquid. Due to the electrochemical reactions between hydrogen/natural gas/coal gas and oxidants, electrons are transferred through the electrolytic liquid, thus resulting in a potential difference between the positive and negative electrodes. The potential difference creates an external voltage; thus, a current flows in the external circuit. The energy production process is illustrated in Figure 9.4.

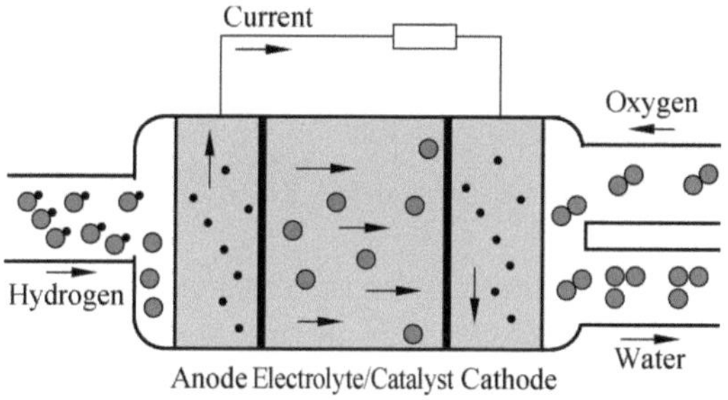

Figure 9.4: Operating principle of a fuel cell.

Based on the types of electrolytic liquids, we can categorize fuel cells into alkaline fuel cells (AFCs), phosphoric acid-type fuel cells (PAFCs), molten carbonate fuel cells (MCFCs), solid oxide fuel cells (SOFCs), and proton exchange membrane fuel cells (PEMFCs) [86]. Based on the temperature under normal working conditions, AFCs (100 °C), PEMFCs (100 °C) and PAFCs (200 °C) are low-temperature fuel cells, while MCFCs (650 °C) and SOFCs (1,000 °C) are high-temperature fuel cells.

Currently, AFCs are widely applied in space technology, and PEMFCs are used as vehicle batteries and small-scale energy supplies. The use of PAFCs as medium-sized energy supplies has been commercialized, and MCFCs have passed industrial tests. Even though SOFCs are a new fuel cell technique, they represent the most promising technique, and a few n × 10 kW SOFC devices have been operated under normal conditions for several thousands of hours.

9.6.1.4 Tidal energy resources

The gravitational force of the moon causes tidal effects that result in periodic rises and declines in sea level. The energy generated from the rising and declining sea level is called tidal energy. There are two tidal energy resource technologies. The first type utilizes geographic landforms to build dams, along which a hydropower plant is sited to produce electricity. Without the construction of a dam, the second type utilizes ocean currents to generate electrical energy. Most operational tidal energy resources belong to the first type. China has eight of the first type of tidal power plants with a total capacity of 6,000 kW and an average annual energy production of 100 GWh.

In 2008, the first of the second type of tidal power plant – Seagon – was installed at the Gulf of Westminster with a capacity of 1.2 MW. Seagon uses ocean currents to produce energy and is the largest of the second type of tidal energy plant. It was also equipped with sonar devices developed by Tritech for five years of continuous monitoring and detection to investigate the impacts of Seagon on nearby marine organisms. In the absence of dam construction, this technology has limited influence on the local marine ecosystem.

One of the most significant challenges is dealing with equipment corrosion and biological attachment problems, since tidal energy plant equipment is immersed in the ocean. It is also important to study and reduce the influences on local marine ecosystems.

9.6.1.5 Biomass power plant

Bioenergy is stored in biomass and is a type of chemical energy converted from solar energy through photosynthesis by chlorophytes. Bioenergy is also a renewable energy and is typically classified into six categories: a. wood and forest industry waste; b. agricultural waste; c. marine plants; d. oil plants; e. urban and industrial organic waste; and f. animal excrement.

Unlike fossil fuels, bioenergy is renewable. Compared to other renewable technologies, bioenergy is relatively adequate and widely distributed and has broad applications (it can be converted into many other types of secondary energy such as electricity generation, heating, and gas/liquid fuel production). In addition, unlike other renewable technologies, bioenergy is manually controllable; thus, biomass energy resources are dispatchable resources, which are very beneficial to the reliable operation of power systems. With the development of biomass energy resource technology, biomass power plants will be an important part of the future power grid. For example, biogas power plants may be a good choice for large pig farms. Furthermore, urban and industrial organic waste is a good illustration of the use of waste products, but the emissions must be treated properly.

9.6.1.6 Gas-fueled cooling–heating–electricity cogeneration

Gas-fueled cooling-heating-electricity cogeneration systems [87] mainly adopt gas turbines, gas internal combustion engines, and gas combustion engines as generators. Waste heat utilization equipment typically includes exhaust heat boilers, vapor absorption refrigeration machines, water absorption refrigeration machines, and exhaust absorption refrigeration machines. We can combine different generators and waste heat utilization equipment to build various gas-fueled cooling-heating-electricity cogeneration systems. The major sources of gas are natural gas and gasified coals. The primary advantages of gas-fueled cooling-heating-electricity cogeneration systems are as follows:

(1) High utilization efficiency

The current overall efficiency in the electricity industry is below 50%. Energy losses become larger as transmission distances increase. If gas-fueled cooling-heating electricity cogeneration systems are installed close to users, electricity production efficiency will be approximately 40%. With the cooling and heating services provided by the utilization of waste heat, the overall efficiency can exceed 80%. In addition, the transmission losses will be reduced by 6% – 7%.

From the energy grade point of view, the electricity from gas-fueled boilers is of low grade, even though they are more efficient than coal plants. However, 35% of the electricity generated by gas-fueled cooling-heating-electricity cogeneration systems is of high grade. Therefore, gas-fueled cooling-heating-electricity cogeneration systems are much more efficient overall than stand-alone gas-fueled boilers.

(2) Peak-shaving and valley-filling effects on gas and electricity markets

In most areas of China, heating is needed during winter, and cooling is needed during summer. The high energy demands during some specific peak periods pose major challenges to the power grid during the summer season. However, heating devices are idle in this period. If gas-fueled cooling-heating-electricity cogeneration systems are adopted, we can use natural gas to provide cooling services and increase gas usage, thereby resulting in less dependence on air conditioners and lower peak loads. Therefore, gas-fueled cooling-heating-electricity cogeneration systems can shave the peaks and fill the valleys for both the electricity and gas markets, by significantly increasing the utilization factors of electricity production and heating devices.

Gas-fueled cooling-heating-electricity cogeneration systems are also very cost-effective and environmentally friendly. Based on reports from the USA, the adoption of distributed gas-fueled cooling-heating-electricity cogeneration systems can reduce maintenance costs and emissions by 12% and 22.75%, respectively for office premises, by 11% and 34.4%, respectively for shopping malls, by 21% and 61.4%, respectively for hospitals, by 32% and 22.7%, respectively for sports centers, and by 23% and 34.3%, respectively, for hotels.

9.6.2 Storage technology

Storage technology converts electricity into the energy of a different type that can be conveniently stored. The stored energy can be converted into another form of energy if needed. Storage technology addresses two issues: how to store massive amounts of energy and how to quickly complete energy conversion [88].

Based on the types of stored energy, we can roughly categorize storage technologies into five types: natural energy storage (such as constructing a dam for storing water upstream and generating electricity – a dispatchable energy resource), mechanical energy storage (such as pumped energy storage, compressed air energy storage, and flywheel energy storage), electrochemical energy storage (sodium-sulfur cells, liquid batteries, and nickel-cadmium cells), electromagnetic energy storage (superconducting magnetic energy storage and supercapacitors) and phase-change energy storage [89, 90].

9.6.2.1 Natural energy storage

Currently, only 23% of all available water resources are used for energy production. We need to advance the development of the most abundant and concentrated renewable natural energy resources to mitigate the delay that has resulted because no large-scale hydropower plant has been approved for construction in the last seven years. The water resources are mostly distributed among the Nujiang River, Jinshajiang River, Yalongjiang River, Minjiang River, Jialingjiang River, Hongshuihe River, Yaluzangbujiang River, and a few other rivers. If all the available water resources are used for electricity generation, the total installed capacity would be 5–6 times higher than that of the Three Gorges Dam, with less stringent immigration requirements for the implementation of those projects. With limitations on developing renewable hydropower plants, it is impossible to achieve an environmentally friendly generation fleet with a significant share of storage devices and renewable energy resources.

9.6.2.2 Mechanical energy storage

Mechanical energy storage devices convert electricity into mechanical energy, such as kinetic and potential energy. These devices consist mainly of pumped energy storage devices, compressed air energy storage devices, flywheel energy storage devices, and so on.

(1) Pumped energy storage devices

Pumped energy storage technology is used for energy storage worldwide, with a total operational capacity of 90 GW, which is almost 3% of the total power plant capacity [91].

When electricity demand is low, pumped energy storage devices use electricity to pump water from a downstream dam to an upstream dam. In contrast, during a high-demand period, the stored energy resources in the upstream dam are released to generate electricity. Based on whether there are natural rivers flowing into the dams, we can categorize pumped storage devices into three groups: pure pumped storage, hybrid pumped storage, and water transfer pumped storage. A fully charged pumped storage device typically needs several hours to several days to be fully discharged. Its energy conversion efficiency ranges from 70% to 85%. Due to the large capacity of pumped storage devices, they are used to manage load curves and implement emergency actions.

The construction of both an upstream dam and a downstream dam requires pumped storage to be sited at specific locations, thereby limiting the development of pumped storage technology. As a typical form of mechanical energy storage technology, it should be possible for pumped storage devices to be installed at all locations satisfying the necessary natural environmental conditions.

(2) Flywheel storage devices

Flywheel storage technology converts electricity into mechanical energy through the kinetic energy of high-inertia wheels driven by a motor. When a flywheel storage device is discharged, the device releases kinetic energy for driving generators to produce electrical energy. A typical flywheel energy storage system consists of high-inertia wheels, a bearing support system, a generator/motor, a converter, a control unit, a vacuum pump, and a few other ancillary devices. The average power density and average energy density for current flywheel storage devices are approximately 5 kW/kg and 20 Wh/kg, respectively. The efficiency of these devices is typically higher than 90%, with a life of 20 years [92].

The main advantages of flywheel storage devices are their long life (approximately 20 years or tens of thousands of full charge/discharge cycles), high energy conversion efficiency (we can further improve the efficiency if a suspension bearing system is adopted), lack of pollution and noise, strong load tracking capability, and environmental friendliness. However, the energy/power densities of these devices are relatively low, and their installation costs are very high. They primarily serve demands that are between short-term and long-term storage applications, including primary frequency management and UPSs.

(3) Compressed air energy storage (CAES) devices

Almost 67% of the fuel in common conventional generators is used to compress air. Consequently, during low-load periods, we can compress massive amounts of air in advance and store it in abandoned mines, submarine gas storage tanks, caves, or oil wells. When the load demands increase, we can extract some compressed gas for electricity generation.

Under the same working conditions and with the same outputs, the gas usage of generation units with compressed air energy storage devices is 40% lesser than that of conventional generators. The first commercialized compressed air energy storage device was installed in Germany in 1978, with a total capacity of 290 MW, and has been in operation for more than 30 years. Large-scale compressed air energy storage devices need to be installed at sites with specific geographic structures. With the development of distribution networks and microgrids, small-scale compressed air energy storage devices with capacities of 8–12 MW are of great interest.

9.6.2.3 Superconducting magnetic energy storage

Superconducting magnetic energy storage (SMES) technology connects superconducting coils with power systems through high-power electronic devices. When there is excess energy in the power system, SMES devices convert electricity into electromagnetic energy. The stored electromagnetic energy will be converted into electrical energy by a rectifier and an inverter to supply power to the grid, when needed. In other words, a SMES device charges the superconducting energy storage magnets in advance and controls real/reactive power exchange with the power grid.

Under superconducting conditions, superconducting coils are in operation without any energy loss; thus, their current density is typically higher than that of common coils by approximately 1–2 orders of magnitude. Therefore, these coils can store energy for a long period with minimal energy loss and a high current density. The possible energy loss comes mainly from temperature maintenance and magnetic flux leakage in the superconducting magnets.

Compared to other energy storage technologies, SMES can realize energy exchange in four quadrants (+ real + reactive, + real – reactive, – real – reactive, and – real + reactive) with a high conversion efficiency, short response time, and flexible operations. Initially, SMES was used to manage load curves. Later, SMES was also applied to provide damping effects, mitigate system electromagnetic oscillation, and improve electricity quality and reliability.

The primary limitation on the development of SMES devices comes from the performance and cost of the superconducting materials. Currently, there are five generations of high-temperature superconducting materials: lanthanide superconductors, Y-system superconductors, bi-system superconductors, thallium superconductors, and mercury system superconductors. Among these materials, Y-system superconductors and bi-system superconductors are the most useful [93]. Currently, there is a significant gap between superconducting magnet performance and the requirements of SMES, and this gap poses challenges to the commercialization of SMES devices.

Due to its fast response capability and lossless energy conversion process, SMES technology will be very popular and useful for the development of SGs, once we significantly improve the performance and reduce the costs of the superconducting materials.

9.6.2.4 Electrochemical energy storage

Two major categories of electrochemical energy storage devices are supercapacitors and batteries. They are described as follows:

(1) Supercapacitors

Supercapacitors, also called electrochemical capacitors, are a storage technology between conventional capacitors and batteries and began to be developed in the 1970s [94].

According to the energy storage mechanism, we can group supercapacitors into electric double-layer capacitors (EDLCs) and faradaic pseudocapacitors. The charge separation procedure in the electrodes and the electrolytic liquid leads to the double-layer capacitance of an EDLC. The capacitance of faradaic pseudocapacitors results not only from the charge separation procedure in the electrodes and the electrolytic liquid but also from the electrons generated during the oxidation-reduction reaction in the electrolytic liquid. Consequently, faradaic pseudocapacitors typically have larger currents. For example, a faradaic pseudocapacitor using RuO_2 has a current 10 – 100 times larger than that of EDLCs.

Compared to traditional capacitors, supercapacitors have a larger capacity, higher power/energy density, faster charge/discharge capability, longer life, and less sensitivity to environmental changes. If the capability of supercapacitors to resist high voltages can be enforced, the performance of supercapacitors can be further improved. Due to the high installation costs of supercapacitors, they are used mainly for short-term, high-output applications, such as start-up support for direct current generators, dynamic voltage restorers, and improvement of post-contingency dynamic performance and voltage levels.

(2) Batteries

A battery is charged/discharged by the currents generated by oxidation-reduction reactions between the positive and negative electrodes. Batteries are one of several widely applied electrical storage systems. The application of storage systems in power systems requires that storage devices have a large capability, high efficiency, long life, and low cost. These requirements limit the application of conventional lead-sulfur batteries, nickel-cadmium cells, and lithium batteries in power systems. Thus, sodium-sulfur cells and liquid batteries are more suitable for large-scale applications in power systems.

Sodium-sulfur cells use Na-beta-Al2O3 as the electrolyte and diaphragm, with sodium and sodium polysulfide at the positive and negative electrode sides, respectively. Theoretically, these batteries can have an energy density of 760 Wh/kg, a charge/discharge efficiency of almost 100%, and overall conversion efficiency of 80%. A sodium-sulfur battery consists of multiple sodium-sulfur cells with a total capacity of thousands of kilowatts. Since sodium-sulfur cells are convenient to transport, store, and install, there are over 100 operational sodium-sulfur battery systems sited in the USA and Japan. Sodium-sulfur battery technology is the most mature and promising technology.

Liquid batteries store the positive and negative active substrates in two different tanks of electrolytes separated by an Ion exchange membrane. The electrolytes containing positive and negative active substrates are transported through the battery by a liquid feed pump. These batteries are chemical energy devices that were introduced by L. H. Thaller in 1974 [95]. There are multiple systems of liquid battery technologies. The electrochemical polarization in a liquid battery is very small, and the deep charge/discharge efficiency is approximately 100%. A liquid battery has a relatively large capacity that can be improved by increasing the number of stored electrolytes or the electrolyte concentration. Since the electrolyte tanks and the cells can be installed separately, we can design a liquid battery system of any size or shape based on local geographic structures.

Currently, sodium-sulfur cells and liquid batteries have been commercialized, but their development is limited by the current state of these technologies and costs. With increasing battery capacities and the development of integration technology, the manufacturing and installation costs can be further reduced. Once long-term reliability and security tests are passed, hourly or daily storage systems can be used to improve the stability of renewable energy resource power outputs, modify load characteristic curves, and provide power in the form of UPSs.

9.6.2.5 Phase change energy storage

Phase change energy storage technology uses the energy exchange during the phase-change processes of a few materials for energy conversion and storage purposes. Based on the phases during energy conversion, we can categorize phase change energy storage devices into four groups: solid-solid, solid-liquid, liquid-gas, and solid-gas. According to the materials used, there are also inorganic, organic, and hybrid phase change energy storage technologies.

Phase change energy storage technology is very useful in the space industry, solar applications, heating and air conditioners, peak load management, waste heat utilization, seasonal heating/cooling management, food preservation, energy-saving applications in buildings, agriculture, and so on.

Phase change energy storage technology is an important part of peak load management approaches for SGs. A storage-based air conditioner technique using phase-change materials is widely used to manage peak shaving and valley filling globally. Generally, the storage-based air conditioner technique uses ice or materials with a transformation temperature higher than 5 °C as phase-change cooling materials. Compared to other peak load management approaches, the storage-based air conditioner technique using ice has a very low investment cost (approximately 1,200 RMB/kW). In addition, the storage-based air conditioner technique also has minimal operational costs and few requirements for location sites. Therefore, the storage-based air conditioner technique shows very large potential to be applied in power systems, on a large-scale. The USA, Japan, and many European countries began to develop and widely promote this technique in the 1980s, while China began to introduce this technique around 1995. For instance, in the Shanghainan railway station, the installed storage-based air conditioners make ice from 10 p.m. to 6 a.m. the next morning. The ice, with cooling energy of 22,500 kWh, is used for cooling and results in a savings of 1.225 million yuan (RMB) per year. The Newmart in Dalian also adopted the storage-based air conditioner technique, and the one-time investment in this technique was

4.2 million yuan (RMB). The capacity of storage-based air conditioners is reduced by 25%. Under full-load operating conditions, the installed storage-based air conditioners manage energy totaling 2,765 kWh. The total energy consumption is reduced by 15,000 kWh, and the peak load is reduced by 2,000 kWh (36% of the peak load). Under half-load operating conditions, the installed storage-based air conditioners manage energy totaling 5,580 kWh. The total energy consumption is reduced by 12,260 kWh, and the peak load is reduced by 1,324 kWh (60% of the peak load). Furthermore, the installation of storage-based air conditioners lowers the mart's operational costs by 11.2%.

In addition, the storage-based air conditioner technique can be applied to utilize the waste heat from generation units by storing the energy using phase change heating materials for heating in the future. For electric heat storage systems, electric boilers heat water and store it in a tank with phase change heating materials during nonpeak load periods. The stored energy in the form of heat will be used to heat the water, instead of using electricity during high-load periods.

However, the technologies of flywheel energy storage to electrochemical energy storage are not comparable to natural energy storage and hydropower plants from both the economic and performance points of view.

9.6.2.6 The largest energy storage system – electric vehicles with double electric power interfaces

Due to the increasing awareness of global warming and environmental problems, electric vehicles are of great interest for investment and development in the auto industry. The integration of many electric vehicles leads to significant operational flexibility in an SG because the integration of electric vehicles with the power system is similar to the incorporation of many distributed energy storage devices, which serve as controllable resources and help mitigate the uncertainty and intermittency of renewable energy outputs.

In August 2008, the Ministry of Industry and Information Technology of the People's Republic of China published the *Energy Saving and New Energy Vehicle Development Plan (2011–2020)*. This plan notes that the new energy vehicle industry and market in China will have the largest scale worldwide, in which the number of new energy vehicles (including plug-in hybrid electric vehicles, blade electric vehicles, and hydrogen-powered vehicles) will be approximately 5 million, and the number of hybrid electric vehicles will be approximately 20 million. In other words, the largest energy storage system in the world will be formed. The typical capacity of the battery system of an electric vehicle (BYD F3e) is 310 V 120 A · h, and the system takes 10 h and 42 kWh to be fully charged. Due to the imperfect energy conversion efficiency, the total energy consumption of the battery system will be approximately 45 kWh. Assuming that there are 20 million such vehicles, the total capacity of the energy storage system would be 900 million kW or, equivalently, five times the capacity of the Three Gorges hydro plant. Even assuming that the efficiency will be as low as 0.2, the energy storage system will still have the same capacity as the Three Gorges hydro plant. Such a large energy system will significantly contribute to the energy balance, smoothing renewable energy power outputs, systemwide optimal controls, and so forth.

During the construction of the largest energy storage system, a decision-maker should pay enough attention to the following topics:

(1) Construction of battery recharging stations and piles for electric vehicles with double electric power interfaces

On April 28, 2010, the National Technical Committee of Auto Standardization approved four national standards – *Conductive Charging Interface for Electric Vehicle*, *General Requirements for Electric Vehicle Charging Stations*, *The Communication Protocol between The Battery Management System of Electric Vehicle And Non-Vehicle Charger*, and *Test Methods for Energy Consumption of Light-Duty Hybrid Electric Vehicles*. The establishment of these four national standards has fostered the construction of a comprehensive set of standards for the new energy vehicle industry and the development of new energy vehicle technology. Specifically, *General Requirements for Electric Vehicle Charging Stations* states that the fundamental functions of charging stations for electric vehicles should include charging, monitoring, and measurement; the advanced functions should cover battery charging, battery inspection, and battery maintenance. *Conductive Charging Interface for Electric Vehicles* provides configurable ways of charging electric vehicles.

From the SG point of view, if the largest energy storage system formed by multiple electric vehicles is applied only for charging, the energy storage system can simply help fill the valleys during low-load periods but cannot help shave the peaks during high-load periods. Compared to the usefulness of hydropower plants with peak-shaving and valley-filling capabilities, the usefulness of the energy storage system formed by multiple electric vehicles would be greatly weakened.

Therefore, the authors suggest that the authorities should establish standards and development plans for constructing battery recharging stations and piles with bidirectional electric power interfaces. Constructing battery charging stations and piles with unidirectional electric power interfaces with later upgrades is a waste of energy and investment.

(2) Disposal issues for used batteries

The massive number of used batteries is a major challenge to be addressed in the future since a piece of button cell requires 0.6 million gallons of water. The average life of a typical electric vehicle battery is approximately 3 to 5 years. The inappropriate disposal of used batteries will negatively affect the environment. We should pay attention to this issue as soon as possible.

A few developed counties have been exploring the issue of battery disposal for years, with some useful results and experiences. For example, in the USA and

Japan, the government will compensate industrial users for the proper disposal of used batteries per ton, thereby encouraging industrial users. The Korean government has set rules that every battery manufacturing plant needs to select a factory and pay a service fee of some amount for battery disposal; a few other countries charge additional taxes on battery manufacturing plants and have reduced/removed the tax burden on used battery disposal factories. However, the integration of used batteries into the power grid for further utilization is another way to solve the used battery disposal problem. In general, fully charged batteries that have only 80% of their nameplate capacity will be retired and will no longer be used for electric vehicles. However, Electrovaya in Canada proposes that these batteries are very useful as energy storage system components. Based on this idea, Electrovaya is working with the University of Manitoba on the integration of multiple used batteries to build a 150 kW energy storage system.

(3) Different choices for charging electric vehicles

The need to construct charging stations limits the rapid development of electric vehicles. Due to immature battery technology, the batteries installed in electric vehicles need to be frequently charged/discharged. Public charging stations are the first choice for charging electric vehicles; thus, these stations have strict requirements for charging speeds and charging durations. The choice of method for charging electric vehicles affects the long-term performance of batteries for these vehicles.

The current charging approaches include the common charging method, quick charging method, and mechanical charging method.

The common charging method is flexible in terms of charging speeds and charging durations. Thus, electric vehicles adopting the common charging method can be regulated and controlled using the EMS mentioned in the previous chapter. As energy storage devices for residential and commercial users, electric vehicle batteries provide substantial flexibility and support to power systems. Electric vehicles also enable users to adjust their energy usage at different times over a day to reduce energy consumption costs. Users must specify settings on the EMS, such that the electric vehicles will be ready to use when needed.

The quick charging method is also called the emergency charging method and charges electric vehicles for a short duration (20 min to 2 h) by using large currents (150–400 A). The quick charging method can complete a full charging process quickly, but batteries adopting this method will have a short life. Compared to the common charging method, the quick charging method has relatively low overall efficiency and negatively influences power systems.

The mechanical charging method is a direct method that "charges" electric vehicles by changing their batteries. Due to the weights of the batteries, the charging process requires professional staff to complete the changing, charging, and maintenance functionalities. This method can complete the "charging" process within several minutes and enables the staff to inspect the unequipped batteries. However, this method requires stations to have the same batteries. This requirement is very hard to satisfy. If public charging stations adopt the mechanical charging method, the batteries connected to the grid in the stations can be used as energy storage plants that benefit the reliable operations of power systems.

9.6.3 Economic interactive energy-consuming techniques

The final goal of energy production is to serve users and provide them with clean, high-quality, reliable, and inexpensive electricity. In conventional power systems, the only role of the users is as electricity consumers, who cannot actively participate in power grid operations. This constraint limits the flexible and economical operations of power systems. An SG highlights the importance of the active participation of users in power system operations and the bidirectional interactions between users and the power grid. Instead of being pure consumers, users, as participants in the power system operation, can design their own energy consumption activities, choose the best electricity service, and control distributed energy resources, based on the current power system status. The power grid can also make the corresponding commitment and dispatch plans. For the power grid, the users serve as a controllable resource, such that the whole power system is fully controllable, thus resulting in significant contributions to the economical and reliable operations of power systems.

Economic interactive energy-consuming techniques have two primary advantages: reducing network energy losses and improving the service quality. To make use of these advantages, we have to install smart appliances, improve advanced measurement systems, and implement demand-side services at the same time. Since appliances are direct energy consumers, the enforcement of smart appliance development benefits the effective utilization of electricity. Constructing a wide area measurement system enables the system operators and end-users to know how the energy is consumed. As a result, consumers can actively make a better electricity usage plan to reduce energy consumption and improve overall efficiency. In addition, demand-side management and time-varying electricity pricing will provide reasonable guidance to foster developing economic interactions between the users and the grid so that the energy losses of the grid are reduced.

9.6.3.1 Smart appliances

Similar to SG concepts, smart appliances continue to change and develop. The first definition of intelligence for home appliances is to apply digital techniques, computer science, and information technology to equip traditional appliances with advanced functions, such as remote controls, remote maintenance, automatic alarming, and automatic upgrades. In addition to the functions mentioned above, smart appliances should also enable interactions with the power grid to adjust electricity consumption,

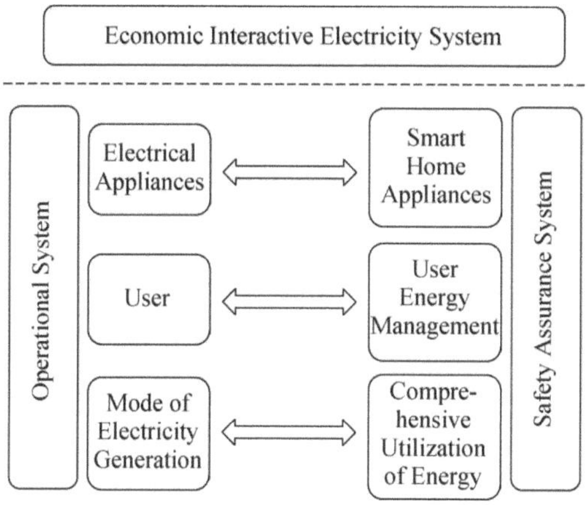

Figure 9.5: Functional framework of an economic interactive energy consumption system.

by using real-time energy prices and system frequencies. By doing so, the end-users can reduce their electricity costs and increase overall efficiency.

Current communications of users, especially residential users, with the power grid are still very weak. We can divide smart appliances into two categories: those that need to interact with the grid and those that do not.

Smart appliances without an immediate need to interact with the grid do not communicate with the grid in real-time. Such appliances include residential heaters, central air conditioners, washers, and dryers. These appliances do not have strict requirements for start-up time. Additionally, short-term interruptions of these appliances have limited impacts on users.

Pacific Northwest National Laboratory (PNNL) started a project to develop grid-friendly appliances in 2006. This project upgraded many appliances, such as heaters and dryers, which could stop working for a predetermined duration (2 min) when the system frequency is below 59.5 Hz (the nominal system frequency in the USA is 60 Hz). If the system frequency is still below 59.5 Hz after a two-minute period, these smart appliances will be forced into an outage for another 2 min. However, the total interruption cannot be more than 10 min. In later studies, the collected information and data verified that this project did not bring any inconvenience to users. In England, RLtec company is in control of a set of refrigerators based on the measured system frequencies that are used to adjust the electricity usage to reduce the system frequency fluctuations.

The aforementioned grid-friendly appliances are simply products in the first phase of smart appliance development. The most significant advantage of the approach developed by PNNL is that it requires only the installation of system frequency measurement modules at the user endpoints, thus resulting in reasonable investment costs. To further reduce the costs, the Econnect company in England has proposed the concept of a smart plug-in, which will disconnect appliances from the grid once the frequency is below a threshold value. Some isolated power systems have already adopted this idea.

Similar ideas can be applied for voltage regulation and reactive power provisioning. When the system voltage level drops, we can temporarily stop appliances to recover the system voltage quality and level. The application of smart appliances to achieve the steady operation of the power system is similar to the automatic approaches used in traditional power systems but does not blindly turn on or turn off all the appliances at the same time. These applications guarantee the fundamental demands for electrical energy and may stop supplying electricity to some lower-priority users when needed.

Further development of smart meters, advanced measurement systems, residential communication networks, and user EMSs will foster interactions between smart appliances and the power grid. The bidirectional information exchanges between users and the power grid enable the users to adjust their electricity consumption activities based on the real-time operational status and electricity price information of the power system. Users can also authorize the system control center to manage and regulate their appliances remotely to reduce their electricity costs or improve the quality of the electricity they receive. Most end users only need to set several parameters to intelligently manage their installed appliances. For example, in commercial buildings and hotels, the managers can set up the parameters for smart appliances: the target water temperature in a heater will be lowered automatically when no one is on-site, or the target temperature set for air conditioners will be increased by 4 degrees during peak load periods. Once the users are willing to do so, the control center can automatically control their participating appliances.

9.6.3.2 Advance metering infrastructure

In a traditional power system, the load is thought of as uncontrollable; thus, the control center cares little about the internal activities at the user endpoints. Therefore, among the generation, transmission, distribution, and demand sides, only the first three are thought to be dispatchable and observable, while the demand side is not.

In a SG, with the introduction of distributed energy resources, demand-side management, and EMSs at the user end, electricity consumers can actively participate in grid activities so that the consumers are controllable and dispatchable as the members are on the generation, transmission, and distribution sides. In this case, we need to develop the necessary monitoring system on the demand side. The end-users need to obtain sufficient information on the real-time electricity prices and system operational status to adjust their energy consumption activities. By doing so, the users can reduce their costs of energy consumption and increase system efficiency.

AMI is a product developed in this context. Implementation of AMI enables bidirectional information exchange between users and the grid; examples of AMI include remote controls, time-varying pricing mechanisms, and demand-side management. Figure 9.6 shows an illustrative representation of AMI. Typically, AMI consists of smart meters at the user endpoints and data collectors in the power companies and the associated communication network [96].

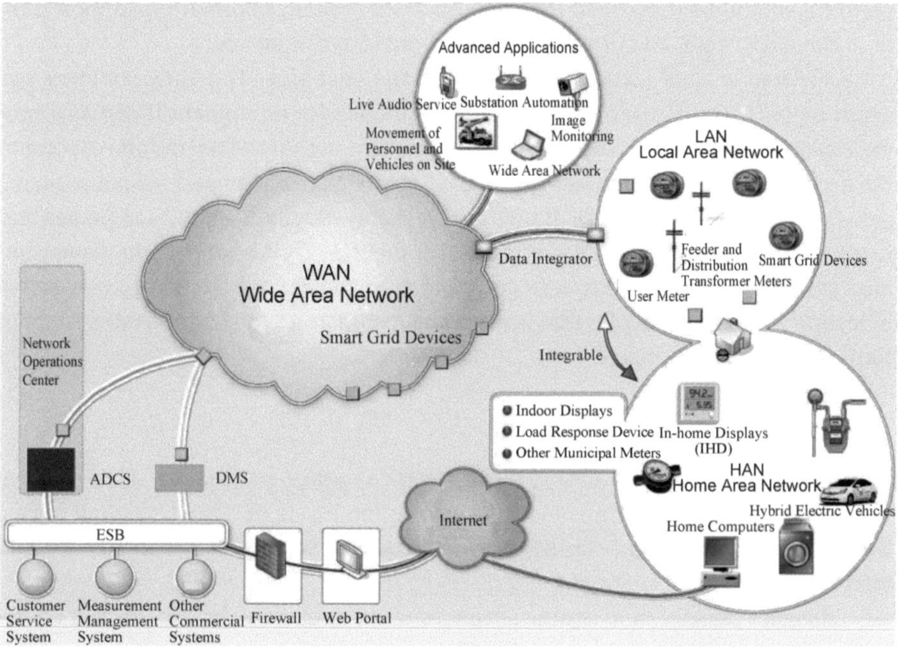

Figure 9.6: Illustrative representation of AMI [98].

Smart meters are the fundamental units of an AMI system. For common residential users, conventional meters only record electricity usage and do not provide any useful information, particularly real-time energy prices. The functionalities of a smart meter are more powerful than those of a conventional meter. A smart meter can be used to measure the current, voltage, harmonics, and so on, in real-time. In addition, a smart meter is equipped with a programmable module for timing measurement, storing measured data, outage alarming, cutting-off loads, and other applications [97]. Another important feature of a smart meter is the communication module, which serves as the gateway between the power system communication network and a home area network (HAN), as follows: 1) the users can receive the dispatch commands from the distribution network control center; 2) the users' real-time energy usage information can be uploaded via the HAN to the local distribution center. By doing so, we can realize bidirectional information flows (streaming) to improve the observability and controllability of the power system.

A data management system is at the core of an AMI system. Constructing AMI requires equipping a data management system in the control center to receive the massive amount of data for collection, management, and utilization. Reports [96, 97] have discussed the functions of a data management system in detail. For planning and constructing a SG, the data management system, and the SG information support platform should be integrated together, such that the SG information support platform can filter, store, and use the data measured by smart meters.

A HAN acts as an extension of AMI on the user side. The HAN connects the smart meter with the users' controllable appliances or equipment (such as a programmable temperature controller) via a gateway so that the users can respond to the needs of the power company and actively participate in demand–response activities. Currently, due to the lack of uniform technical standards, many companies and institutions, such as Homeplug, Ethernet, HomePNA, ZigBee, and Upnp, have commercialized their own products. In addition, many power companies are very active in testing AMI. For example, IBM is planning to construct the largest AMI system in North America and Europe.

9.6.3.3 Demand-side management

Demand-side management aims to direct electricity usage activities for more effective utilization of energy and transmission/distribution assets by adjusting retail prices. Demand-side management consists of two primary parts, efficiency management and load management:

(1) Efficiency management adopts advanced technologies and highly efficient equipment to improve terminal energy efficiency and reduce electricity losses to save energy and reduce emissions. In addition, efficiency management can modify the system load characteristic curve by shaving the peaks and filling the valleys, thereby leading to many social and economic benefits.
(2) Load management improves users' energy usage behavior by modifying the system load characteristic curve by shaving the peaks and filling the valleys.

There are two kinds of natural imbalances in power system loads: space and time. These two types of imbalances result in the need for demand-side management. A time imbalance refers to the varying time when the load reaches its peak or valley points; a space imbalance indicates the unbalanced geographic distribution of the load and generation due to resource constraints and economic development limits. These imbalances increase investment in transmission and distribution networks as also the overloading of equipment during peak-load periods. Under the overload conditions of transmission/distribution networks, it is not possible to fully utilize some generation.

From an economic point of view, demand-side management requires power companies to determine electricity prices based on the locational and time-dependent values of electric energy. The time- and location-varying prices will direct the users' energy usage activities and consumption. Implementing these functions requires support from power market policies and communication techniques. The multiple advantages and very large contributions of demand-side management to energy savings and emission reduction underline the importance of demand-side management for an SG:

(1) We can commit and dispatch distributed energy resources at the user end by using demand-side management technology. Based on the real-time operational status of the power system and electricity prices, we can also adjust the production plan of these distributed energy resources and microgrids.
(2) With the installation of energy storage devices, we can shave the peak loads or fill the valleys according to users' electricity consumption and the system's peak/valley distributions.

9.6.4 New operation and control methods for distribution networks

9.6.4.1 Virtual power plants

With the rapid development of distributed energy resources, the number of generation units connected to the power grid will soon increase sharply, thus imposing new demands on the distribution network control methods implemented by the control center. In addition, due to the relatively small capacities and the diverse fuel types of distributed energy resources, these resources have various output characteristics. For example, wind farm outputs are typically higher during the night and lower in the daytime, while the energy produced by solar power plants depends heavily on the weather and solar irradiation. The availability of fuels and rainfall levels greatly affect fossil fuel power plants and small-scale hydro plants, respectively. Due to these distinctive features and characteristics, the controllability of distributed energy resources of a single type is limited, thus negatively affecting the effective utilization of distributed generation units. Hence, the Flexible Electricity Networks to Integrate the Expected Energy Evolution (FENIX) program in Europe provides the definition and concepts of a virtual power plant (VPP). A VPP refers to a set of controllable loads and distributed energy resources that can be controlled and dispatched by the system control center. While doing so, the control center and independent system operators need not know the detailed information of each distributed generation unit. However, they need to communicate with, control, commit, and dispatch the VPP, which will accordingly adjust the outputs of the distributed energy resources and loads. The introduction of a VPP reduces the burden on the grid control centers and independent system operators. Furthermore, the complementary effects among different types of distributed energy resources make the VPP easily controllable and dispatchable by the control center.

In the FENIX program, technical VPPs (TVPPs) and commercial VPPs (CVPPs) are the two primary categories of VPPs. A single distributed energy resource may belong to a CVPP and a TVPP at the same time.

A TVPP consists of distributed generators within one area. A TVPP analyzes the generation costs and operational status of local distributed-generators and constructs an appropriate VPP model that can be used to reflect the properties of combinations. Additionally, the TVPP provides a few ancillary services to the grid. A TVPP always requires information on the local distribution networks and their associated management system.

A CVPP contains the generation costs and operational status of the distributed generation members in the CVPP. In contrast to a TVPP, a CVPP does not consider the impacts of distribution networks. The main goal of a CVPP is to effectively integrate distributed energy resources with small capacities for participation in the electricity market.

Instead of completely updating the generation fleet and the power system operation patterns, the introduction of VPP concepts aims to integrate the available energy resources. For the generation side, the VPP conceptually combines many distributed energy resources, and a VPP is built to be committed and dispatched by the control center. Such an idea reduces the uncontrollability of wind and solar power, effectively increases the stability of system operations, and improves the electricity grades. Additionally, the owners of these distributed energy resources can choose some of the energy resources for energy production and participate in energy trading activities via the VPP, based on real-time system information. Therefore, the implementation of a VPP can increase the revenues of those owners.

Currently, the FENIX program has built two VPPs in Woking, England (North Program), and Alava, Spain (South Program). The North Program focuses mainly on investigating the values of a CVPP in electricity markets, such as residential photovoltaic (PV) solar plants and small-scale combined heat and power (CHP) plants connected to low-voltage networks; the South Program focuses primarily on the capability of distributed energy resources integrated into medium-voltage networks to provide ancillary services to transmission system operators (TSOs) and distribution system operators (DSOs). FENIX program staff currently use the real data measured in these two VPPs to perform the studies mentioned above.

In addition to the FENIX program, some companies also actively develop experimental VPPs. The framework of a VPP developed by IBM in Denmark is shown in Figure 9.7. This VPP program is designed to include 1% of the national electricity capacity to realize the control and optimization of renewable energy resources.

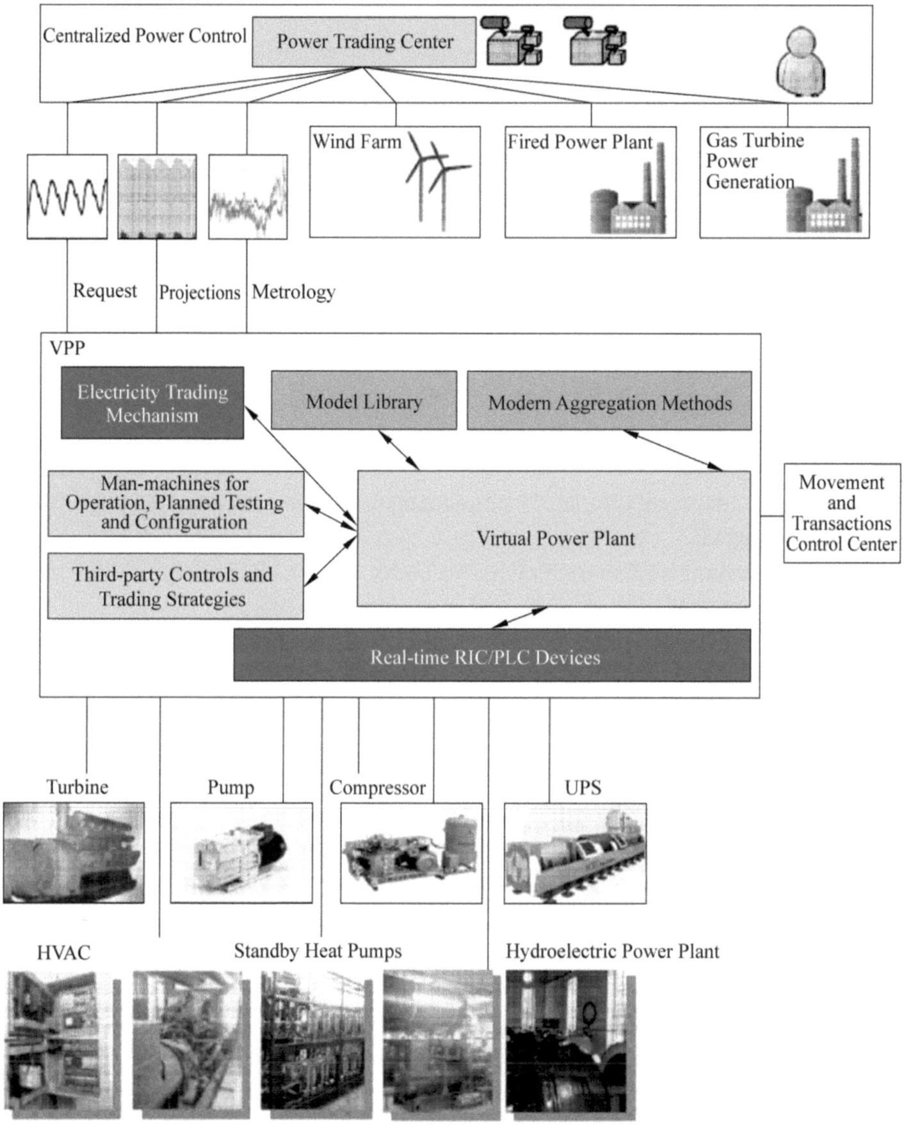

Figure 9.7: The framework of a VPP developed by IBM in Denmark.

9.6.4.2 Microgrids

On the one hand, since distributed energy resources are more flexible, closer to load centers, and require lower-cost construction of transmission/distribution lines, the integration of distributed energy resources improves the reliability of the power supply. On the other hand, the deepening penetration of massive distributed energy resources significantly affects the power grid. Due to the high integration costs and limited controllability of every single plant, it is challenging to manage and dispatch distributed energy resources.

Therefore, a few developed countries have proposed a new concept – a microgrid: generation units, loads, and corresponding distribution networks form a relatively independent small-scale power grid. For example, high-priority loads are satisfied in the distribution network by building a power plant. Such a distribution network is called a microgrid. The microgrid exchanges energy with external power systems through the power conditioning unit (PCU). Currently, there is no uniform definition of a microgrid. Work [99] summarizes some important work on the definition of a microgrid:

(1) The Consortium for Electric Reliability Technology (CERTS) defines a microgrid as a system consisting of loads and micro supplies that can provide both electricity and thermal energy. The power electronic devices in the internal supplies of a microgrid are the primary components that convert energy and provide the necessary controls. For a large-scale external system, a microgrid can be viewed as a single controllable module that can satisfy the local electricity demand and requirements for energy quality and supply reliability.

(2) The European Commission Project Microgrids notes that the fundamental features of a microgrid should include the utilization of primary energy, the application of a small-scale power supply to build a cooling-heating-electricity generation unit, the incorporation of storage devices, and the use of power electronics for control.

(3) R. H. Lasseter from the University of Wisconsin–Madison gives the following definition: a microgrid is an independent, controllable system that consists of loads and small-scale power supplies to provide electricity and heating services. This definition introduces a new model to describe the operations of a microgrid: a microgrid can be regarded as a controllable unit in a large-scale power system; a microgrid can respond to external demands for electricity within seconds and satisfy some specific needs of users; and a microgrid can also improve the reliability of the local network, reduce energy losses, maintain voltage levels, utilize waste heat to increase the overall energy conversion efficiency, and provide uninterrupted electricity to loads.

Based on the definitions given above, a microgrid aims to integrate generation units, loads, storage devices, and controllers together to build a relatively independent, controllable unit that can provide both electricity and heating services. In addition, a microgrid is regulated by the power system control center. In recent studies on microgrids, the final target is to build a microgrid as a controllable cell of the power system; thus, the microgrid can quickly respond to commands from the power system control center to satisfy the requirements. For users, a microgrid can provide the corresponding services based on real-time needs, such as improving local system reliability, preventing sudden voltage drops, and reducing network losses. Satisfying the various requirements of the power system and providing diverse services to users are two important features of microgrids [100]. Since a microgrid is connected to a large-scale power system purely via the PCU, the integration standards for microgrids are not specified for any individual micro power supply. Thus, it is better to consider only the integration issues of one microgrid instead of those of multiple distributed energy resources, loads, and so on.

Since both microgrids and VPPs are solutions proposed for the integration problems of distributed energy resources, we are also curious about the differences between these solutions:

(1) From the point of view of the geographic distribution of distributed energy resources, a VPP virtually combines all distributed generation units into a single power plant for dispatch and management. These distributed generation units may not necessarily be located close to each other. In contrast, the distributed energy resources and loads in a microgrid are from a local area.
(2) From the point of view of resource types, a VPP considers only the distributed generation units, while a microgrid includes not only the distributed energy resources but also the local loads and storage devices.
(3) From the point of view of the relationship with the power grid, both VPPs and microgrids are individual, independent, controllable resources. However, unlike VPPs, microgrids can be in operation even when they are disconnected from the power grid under some scenarios.
(4) From the research point of view, VPPs care more about the economic aspects, while microgrids pay more attention to the users' electricity quality and power supply reliability.

In summary, a microgrid is similar to a small residential community that can function as a power grid, but a VPP is simply a power plant with one single function. Therefore, in constructing an SG, the users and power companies can choose whether they need to adopt a microgrid or a VPP as an effective solution for the integration issues of distributed energy resources, according to their own situations.

9.7 Conclusions

The distribution networks and the user side in an SPS can be called a smart distribution network or SG. This chapter defines an SG, discusses ideas for constructing SGs, and introduces up-to-date techniques related to SGs in academia and industry. These new techniques include distributed energy resource technologies, storage technologies, economical energy consumption technologies, and distribution network control technologies.

SGs should be an intensely discussed topic in both academia and industry because the development of SGs significantly fosters related upgrades in industry and innovation in techniques. One more important reason is that constructing SGs can benefit everyone and enhance the sustainable development of the world economy. As an overview of SG concepts and techniques, this chapter focuses primarily on two aspects. First, we must change our approach. Demand-side management should have a higher priority for power system control and management. Additionally, developing an appropriate real-time electricity pricing mechanism and utilizing market rules can further improve the economics of the power system, reduce energy losses, and lower greenhouse gas emissions. Second, we must adopt a closed-loop control strategy, develop new techniques, improve the automation levels of distribution networks, and enrich regulation approaches to achieve multi-index optimum operation. In addition, there is an acute need to build SEMSs for distribution grids to provide online simulation and analysis, optimize decision-making, and realize closed-loop control. By doing so, an SG can stay in a secure, economic, and high-quality, multi-index, optimal, operation state. Currently, we are actively working in these directions and hope to verify their correctness.

Finally, we appreciate that researchers and engineers in the power and energy fields can work together to support the development and application of SG techniques.

References

[1] L U Qiang. Ditital Power System(DPS). Automation of Electric Power Systems, 2000, 24(9): 1–4.

[2] S Callahan. Meeting Future Utility Operating Challenges With a Smart Grid. 2008. http://www-03.ibm.com/industries/global/files/meeting_future_utility_operating_challenges_with_a_smart_grid.pdf.

[3] X Kai, L Yong-qi, Z Zhi-zhong, Y U Er-keng. The vision of future smart grid. Electric Power, 2008, 41(6): 19–22.

[4] Y U Yi-xin, L Wen-peng. Smart grid. Power System and Clean Energy, 2008, 25(1): 7–11.

[5] L Ming, H E Guang-yu, S Chen, L U Qiang. Brief introduction to thr IECSA project. Automation of Electric Power Systems, 2006, 30(13): 99–104.

[6] L I Wei, Z Lei. Development of the PJM operation management and advanced control center. East China Electric Power, 2009, 37(6): 904–906.

[7] IBM. Innovative Operation Management of Building Smart Grid-New Ideas for China's Power Industry, 2006.

[8] EPRI. Advanced Distribution Automation Program Overview. 2008. http://mydocs.epri.com/docs/Portfolio/PDF/2008_P124.pdf.

[9] J P Paul, J Y Leost, J M Tesseron. Survey of the secondary voltage control in France: Present realization and investigations. IEEE Transactions on Power Systems, 1987, 2(2): 505–511.

[10] J L Sancha, J L Fernandez, A C Ortes. Secondary voltage control: Analysis, solutions and simulation results for the spanish transmission system. IEEE Transactions on Power Systems, 1996, 11(2): 630–638.

[11] L U Wei. The voltage control system based on the global method. Zhejiang Electric Power, 2005(05): 56–59.

[12] H L Smith. Substation automation problems and possibilities. Computer Applications in Power, IEEE, 1996, 9(4): 33–36.

[13] L Andersson, C Brunner, F Engler. Substation automation based on IEC 61850 with new process-close technologies. in: Power Tech Conference Proceedings, 2003 IEEE Bologna. 2003.

[14] M Kezunovic. Future trends in protective relaying, substation automation, testing and related standardization. in: Transmission and Distribution Conference and Exhibition 2002: Asia Pacific. IEEE/PES. 2002.

[15] M Schumacher, C Hoga, J Schmid. Future digital substation with all signals via one digital platform. in: Power Engineering Society General Meeting. Tampa, FL, USA, 2007.

[16] W J Dahlgren. Digital Substation based on IEC 61850: Its Features and Benefits for Power System Protection.

[17] T Hillers, O Beierl. Control, Monitoring and Diagnostics for High Voltage GIS, in IEE Colloquium on GIS (Gas-Insulated Switchgear) at Transmission and Distribution Voltages. Nottingham, UK, 1995.

[18] R Yan-ming, Q Li-jun, Y Qi-xun. Study on IEC61850 communication protocol architecture. Automation of Electric Power Systems, 2000(08): 62–64.

[19] IEC. 61850: Communication Networks and Systems in Substations. 2004.

[20] L Jian-min, Z Zheng-yi. Discussion on the transition from planned maintenance to state maintenance. Electric Safety Technology, 2008(06): 60–61.

[21] Q Jic-heng, W U Juan. Application research of risk management system based on condition assessment of power grid. Electric Power, 2007, 40(004): 90–92.

[22] H Zi-liang. Spreed application of fieldbus system overall realize digitization of fossil-fired power plant. Electric Power, 2004(03): 76–79.

https://doi.org/10.1515/9783110448825-010

[23] H Xiao-xia. Discussion on the construction of digital thermal power plant. Electric Power Information and Communication Technology, 2005(08): 33–35.
[24] Z Su-hua. The technical consideration on constructing Chinese digitization power plants. Huadian Technology, 2008(07): 32–36.
[25] H Wen-long. Summary of the development status of digital construction of power plants. Pearl River, 2008(05): 58–60.
[26] W Mei-shu. Large Thermal Power Plant Supervisory Information System (SIS) Design and Implementation. College of Electrical Engineering Zhejiang University, 2007.
[27] Y Jing-yu, Z Wei-chang. A digitized power plant solution and its application. Electric Power, 2003, 36(12): 34–37.
[28] W Xiao-xi. Application of condition monitoring technologies for UHVAC transmission lines. Power System Technology, 2007, 31(022): 7–11.
[29] M A Yi, W Ke. Analysis of lightning parameters in Yunnan Province based on monitoring data of lightning positioning system. Yunnan Electric Power, 2009(02): 4–6.
[30] C Shui-ming, F Ling-meng, H E Hong-ming, Z Ding-zhu. Performance of lighning location system in Guangdong Province. Electric Power, 2001(12): 43–47.
[31] J Wang, Y Wang, S Du, W Wang, F Guo. Research and application of on-line monitoring system of polluted insulations in HV transmission line. Power System Technology, 2007(S2): 18–21.
[32] X Huo, G Wu. An intelligent wires temperature measurement system based on GPrs technology. Electric Engineering, 2005(6): 24–25.
[33] L Tang, J Qiu. A transmission line management system based on three-dimension geographic information system. Power System Technology, 2003, 27(010): 43–47.
[34] X Guan, T Zhang. Condition-based maintenance of transmission line based on related lifecycle information. Journal of Shanghai University of Electric Power, 2008(01): 1–3, 8.
[35] IEC. 61400 Wind turbine generator systems. 2008.
[36] IEC. 61850-7-420: Communications systems for distributed energy resources (DER) – Logical nodes.
[37] IEC. 61400-25: Communications for monitoring and control of wind power plants. 2006.
[38] IEC. 62270:Hydroelectricpower plant automation – Guide for computer-based control. 2004.
[39] IEC. 61968: Application Integration at Electric Utilities – System Interfaces for Distribution Management. 2003.
[40] IEC. 61970 Energy Management System Application Program Interface (EMS-API). 2003.
[41] IEC. 61850-7-410: Hydroelectric power plants – Communication for monitoring and control.
[42] Q Liu, Y Ni. A brief review of power system stability and control. Proceedings of the CSEE, 1990, 10(6): 1–12.
[43] NERC. Performance Standards Reference Document. 2002. http://www.nerc.com/pub/sys/all_updl/oc/opman/PerformStdsRef.pdf.
[44] J P Paul, J Y Leost, J M Tesseron. Survey of the secondary voltage control in France: Present realization and investigations. IEEE Transactions on Power Systems, 1987, 2(2): 505–511.
[45] Q Guo, H Sun, B Zhang, et al. Research and development of AVC system for Jiangsu power networks. Automation of Electric Power Systems, 2004, 28(22): 33–37.
[46] H Vu, P Pruvot, C Launay, Y Harmand. An Improved voltage control on large-scale power system. IEEE Transactions on Power Systems, 1996, 11(3): 1295–1303.
[47] H Sun, Q Guo, B Zhang. Research and prospects for automatic voltage control techniques in large-scale power grids. Journal of Electric Power Science and Technology, 2007, 22(1): 7–12.
[48] B Zhang, S Chen. Advanced Power Network Analysis. Beijing: Tsinghua University Press, 1996.
[49] N Jaleeli, L S VanSlyck, D N Ewart, L H Fink, A G Hoffmann. Understanding automatic generation control. IEEE Transactions on Power Systems, 1992, 7(3): 1106–1122.

[50] F C Schweppe, J Wildes. Power system static-state estimation. Part I: Exact model. Power apparatus and systems. IEEE Transactions on Power Systems, 1970, PAS-89(1): 120–125.

[51] F C Schweppe, D B Rom. Power system static-state estimation. Part II: Approximate model. Power apparatus and systems. IEEE Transactions on Power Systems, 1970, PAS-89(1): 125–130.

[52] R Baldick, K A Clements, Z Pinjo-Dzigal, P W Davis. Implementing nonquadratic objective functions for state estimation and bad data rejection. IEEE Transactions on Power Systems, 1997, 12(1): 376–382.

[53] M K Celik, A Abur. A robust WLAV state estimator using transformations. IEEE Transactions on Power Systems, 1992, 7(1): 106–113.

[54] H Singh, F L Alvarado, W H Liu. Constrained LAV state estimation using penalty functions. IEEE Transactions on Power Systems, 1997, 12(1): 383–388.

[55] L Mili, M G Cheniae, P J Rousseeuw. Robust state estimation of electric power systems. Circuits and Systems I: Fundamental theory and applications. IEEE Transactions on Power Systems, 1994, 41(5): 349–358.

[56] B Li, Y Xue, J Gu, Z Hang. Status quo and prospect of power system state estimation. Automation of Electric Power Systems, 1998, 22(11): 53–60.

[57] V H Quintana, G L Torres. Introduction to interior point methods. in: IEEE PICA, 1999.

[58] R Fletcher, S Leyffer. Nonlinear programming without a penalty function. 1997.

[59] W Hua, H Sasaki, J Kubokawa, R Yokoyama. An interior point nonlinear programming for optimal power flow problems with a novel data structure. IEEE Transactions on Power Systems, 1998, 13(3): 870–877.

[60] E Yu, G Liu, J Zhou. Energy Management System (EMS). Beijing: Science Press, 1998.

[61] Y Sun, G He, S Mei. A new optimal power flow algorithm based on filter interior point method. Advanced Technology of Electrical Engineering and Energy, 2007, 29–33: 53.

[62] M Ulbrich, S Ulbrich, L N Vicente. A globally convergent primal-dual interior-point filter method for nonlinear programming. Mathematical Programming, 2004, 100(2): 379–410.

[63] D A Wiegmann, G R Essenberg, T J Overbye, Y Sun. Human factor aspects of power system flow animation. IEEE Transactions on Power Systems, 2005, 20(3): 1233–1240.

[64] D A Wiegmann, T J Overbye, S M Hoppe, G R Essenberg, S Yan. Human factors aspects of three-dimensional visualization of power system information. in: Power Engineering Society General Meeting, IEEE, 2006.

[65] D A Wiegmann, A M Rich, T J Overbye, Y Sun. Human factors aspects of power system voltage visualizations. in: Proceedings of the 35th Annual Hawaii International Conference on System Sciences HICSS, 2002.

[66] Y Li, J Yao, H Huang, Y Cao, T Han, X Song. Application of SVG in the dispatching automation system of power network. Automation of Electric Power Systems, 2005(23): 80–82.

[67] Q Yuan, K Wang, J Zhang, X Xia. Common information interaction platform based on SVG. Relay, 2005(12): 66–68.

[68] L A Markov, J R Fox, J H Blank. Optimization techniques for two-dimensional placement. in: 21st Conference on Design Automation, 1984.

[69] S Lu, R W Dutton. An analytical algorithm for placement of arbitrarily sized rectangular blocks. in: 22nd Conference on Design Automation, 1985.

[70] J H Hoel. Some variations of Lee's Algorithm. IEEE Transactions on Computers, 1976, C-25(1): 19–24.

[71] C Y Lee. An algorithm for path connections and its applications. IEEE Transactions on Electronic Computers, 1961, EC-10(3): 346–365.

[72] S M Sait, H Youssef. VLSI Physical Design Automation: Theory and Practice. McGRAW-Hill Book Company, 1995.

[73] D Klump, T Overbye. Advanced visualization platform for real-time power system operations. in: 14th Power Systems Computation Conference (PSCC). Sevilla, Spain, 2002.
[74] Y S Ong, H B Gooi, C K Chan. Algorithms for automatic generation of one-line diagrams. IEE Proceedings- Generation. Transmission and Distribution, 2000, 147(5): 292–298.
[75] B Qiu, H B Gooi. Web-based SCADA display systems (WSDS) for access via Internet. IEEE Transactions on Power Systems, 2000, 15(2): 681–686.
[76] J Qiu, Y Qian, Y Liu. Visualization of power system static security assessment based on GIS. Proceedings of the CSEE, 1999(05): 62–66, 84.
[77] W Chen, Y Luo, G Tu, G Sheng. Power system status visualization for dynamic monitoring, Automation of Electric Power Systems, 2004(08): 68–71.
[78] T J Overbye, D A Wiegmann, A M Rich, S Yan. Human factors aspects of power system voltage contour visualizations. IEEE Transactions on Power Systems, 2003, 18(1): 76–82.
[79] Z Hu, J Qiu, K Wang. Implementation of contour visualization for the nodal operation data of power systems. Automation of Electric Power Systems, 2005(08): 55–59.
[80] W Jie, W U Qin, F Shu-li, W Kang-yuan. Implementation and application of voltage contour based on Arc/info. Relay, 2007(09): 65–59.
[81] J D Weber, T J Overbye. Voltage contours for power system visualization. IEEE Transactions on Power Systems, 2000, 15(1): 404–409.
[82] C Jia, S Hong-bin, T Lei. Three-dimensional visualization technique for power system controlcenters and its real-time applications[J]. Automation of Electric Power Systems, 2008, 32(6): 20–24.
[83] F Maghsoodlou, R Masiello, T Ray. Energy management systems. IEEE Power and Energy Magazine, 2004, 2(5): 49–57.
[84] H Ming, Y I Bao-Lian. Progress of fuel cell technology[J]. Journal of Power Sources, 2008(10): 649–654.
[85] C Hai-hong, T U Yi-feng, C Yang. Research status of fuel cell[J]. Journal of Battery Industry, 2003, 8(5): 229–231.
[86] Z Ya-jie, S Xing-wen. Window of Energy World. Beijing: Tsinghua University Press, 2001.
[87] L Xiao-xia, Q U Rui. Research and application of fuel cells[J]. Guangzhou Chemical, 2007, 35(4): 21–23, 26.
[88] C Lin-xin, T Yan-fen, W Jian. Analyzing and exploring of energy consuming in CHP with natural gas[J]. Energy Conservation and Environment Protection, 2005, 4: 5–8.
[89] C Shi-jie, L I Gang, S Hai-shun, W Jin-yu. Application and prospect of energy storage in electrical engineering[J]. Power Grid and Clean Energy, 2009, 25(2): 1–8.
[90] Y U Zhen-hua. Status and development of large capacity energy storage technology[J]. China Power Enterprise Management, 2009(7): 26–28.
[91] Z Wen-liang, Q Ming, L Xiao-kang. Application of energy storage technologies in power grids [J]. Power System Technology, 2008, 32(7): 1–9.
[92] C Shi-jie, W Jin-yu, S Hai-shun. Application of power energy storage techniques in the modern power system[J]. Electrotechnical Application, 2005, 24(4): 1–8, 19.
[93] P G Rutberg, R B Goncharenko, E G Kasharsky, A A Safronov, N G Khanlarova. About prospects of application of the flywheel stabilizer of frequency in a power system. in: Pulsed Power Plasma Science, PPPS-2001. Digest of Technical Papers. 2001.
[94] H Bing-Lin, Z Xue-Wu. New progress of the high temperature superconductor magnetic energy storage[J]. Cryogenics and Superconductivity, 2005, 33(3): 46–50, 54.
[95] B E Conway. Electrochemical Supercapacitors: Scientific Fundamentals and Technological Applications. New York: Kluwer Academic, 1999.
[96] L H Thaller. Electrically Rechargeable Redox Flow Cells. 1976.

[97] L Wen-peng. Advanced metering infrastructure. Southern Power System Technology[J], 2009, 3(2): 6–10.
[98] The National Energy Technology Laboratory for the U.S. Department of Energy. ADVANCED METERING INFRASTRUCTURE, 2008.
[99] S Kun, K Li, Q I Zhi-ping, P Wei, W U Han, X I Peng. A survey on research of microgrid – a new power system[J]. Relay, 2007(12): 75–81.
[100] L U Zong-xiang, W Cai-xia, M Yong, Z Shuang-xi, L Jin-xiang, W Yun-bo. Overview on microgrid research[J]. Automation of Electric Power Systems, 2007, 31(19): 100–107.

Index

https://doi.org/10.1515/9783110448825-011

Bei Fragen zur Produktsicherheit wenden Sie sich bitte an:
If you have any questions regarding product safety,
please contact:

Walter de Gruyter GmbH
Genthiner Straße 13
10785 Berlin
productsafety@degruyterbrill.com